Mitchell Symons was born in London and educated at Mill Hill School and the LSE, where he studied law. Since leaving BBC TV, where he was a researcher and then a director, he has worked as a writer, broadcaster and journalist. He was a principal writer of early editions of the board game Trivial Pursuit and has devised many television formats. He is also the author of more than thirty books, and currently writes a weekly column for the *Sunday Express*.

www.grossbooks.co.uk
www.**rbooks**.co.uk

Why you need a passport when you're going to PUKE
And more crazy-but-true facts!
MITCHELL SYMONS
WINNER OF THE BLUE PETER BEST BOOK WITH FACTS
RED FOX

WHY YOU NEED A PASSPORT WHEN YOU'RE GOING TO PUKE
A RED FOX BOOK 978 1 862 30758 2

First published in Great Britain by Doubleday,
an imprint of Random House Children's Books
A Random House Group Company

Doubleday edition published 2010
Red Fox edition published 2011

1 3 5 7 9 10 8 6 4 2

The Random House Group Limited supports the Forest Stewardship Council (FSC), the leading international forest certification organization. All our titles that are printed on Greenpeace-approved FSC-certified paper carry the FSC logo. Our paper procurement policy can be found at www.rbooks.co.uk/environment.

Set in Optima 12/14.4pt
Red Fox Books are published by Random House Children's Books,
61–63 Uxbridge Road, London W5 5SA

www.kidsatrandomhouse.co.uk
www.rbooks.co.uk

Addresses for companies within The Random House Group Limited can be found at: www.randomhouse.co.uk/offices.htm

THE RANDOM HOUSE GROUP Limited Reg. No. 954009

A CIP catalogue record for this book is available from the British Library.

Printed and bound in Great Britain by CPI Bookmarque, Croydon

To all my friends, family and readers
from around the world

Introduction

Regular readers will have got the drill by now. Tons of amazing random facts, all put together in a colourful-looking book with a very naughty title. Thanks to Barry O'Donovan and James McCann for coming up with this title! Because that's what this book is: a geography book with all the boring bits taken out written by a man who failed every geography exam he ever took (yes, even the multiple-choice ones where you can usually guess your way to 20 per cent) because he was so bored by it all. You can read it from cover to cover (and I'm grateful to anyone who does just that) or else, like most people, you can dip into it for a few minutes at a time. Hopefully, most times you'll find something to justify the investment of your time.
In that sense, it's no different from *How to Avoid a Wombat's Bum, How Much Poo Does an Elephant Do?* and *Why Does Ear Wax Taste So Gross?*: loads of fascinating facts and lists – with the difference that they're all geographically themed.

If only I'd known what fun geography could be, perhaps I'd have become a geographer or even an explorer (albeit the sort who stays in nice hotels).

Now for some important acknowledgements. The first person I want to thank is my son Charlie: without his invaluable groundwork, I really don't know how I'd have even started the book. Other really REALLY important people to mention: (in alphabetical order): my brilliant editor Lauren Buckland, Annie Eaton, Nikalas Catlow and Penny Symons. In addition, I'd also like to thank the following people for their help, contributions and/or support: Gilly Adams, Luigi Bonomi, Paul Donnelley, Jonathan Fingerhut, Jenny Garrison, Bryn Musson, Nicholas Ridge, Mari Roberts, Jack Symons, Louise Symons, Martin Townsend and Rob Woolley.

If I've missed anyone out, then please know that – as with any mistakes in the book – it is, as ever, entirely down to my own stupidity.

Mitchell Symons
www.mitchellsymons.co.uk

That's surprising! (1)

A teenager in India has an unusual party trick. He can drink milk through his nose and squirt it out of his eyes through his tear ducts. Don't try this at home . . .

Only 8.5 per cent of all Alaskans are Eskimos.

The Dogon people use fried onions as a perfume. They rub them all over their body because they think the smell's so attractive.

In France, it's possible to marry a deceased person with the authorization of the President of the Republic – but only in exceptional cases.

Saparmurat Niyazov, the President of Turkmenistan from 1990 until his death in 2006, declared that those who read *Ruhnama, the Book of the Soul* (his book of moral and spiritual guidance) three times a day would automatically go to heaven.

Saparmurat Niyazov was better known as President Turkmenbashi. He changed the calendar so the name of January became Turkmenbashi. And he had a melon named after him. That's right, a melon.

La Marseillaise, France's national anthem, was composed in Strasbourg (in 1792) and not, as you might think, in Marseilles.

The Indus River, from which India derived its name, is now entirely in the territory of Pakistan.

There is a street in Italy that is less than half a metre wide.

In Qatar, the weekend is Thursday and Friday rather than Saturday and Sunday.

A Belgian student couldn't afford a party to celebrate his 20th birthday so he had the

bright idea of offering his friends' foreheads for hire to advertisers on an internet auction site. A company paid them all to have its logo painted on their foreheads for the night of the party.

There is a house in Massachusetts US, which is made entirely from newspapers. The floors, walls, even the furniture is made from newspaper.

Another house, this one in Canada, is made of 18,000 discarded glass bottles.

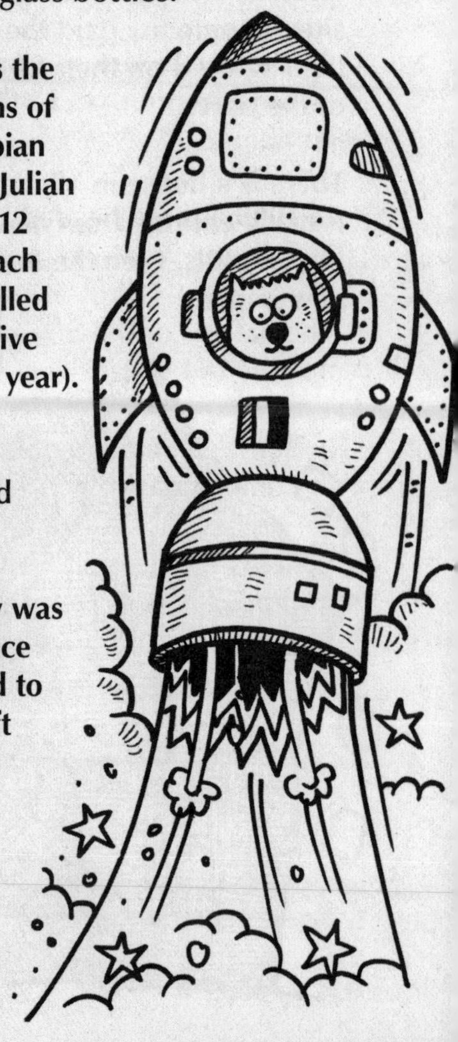

Ethiopia is famous as the country of '13 months of sunshine'. The Ethiopian year is based on the Julian calendar, which has 12 months of 30 days each and a 13th month called Pagume, which has five days (or six in a leap year).

In 1963 the French launched a cat called Feliette into space.

The Statue of Liberty was actually built in France and then transported to America. It was a gift from the French to the Americans.

Bird droppings are the chief export of Nauru.

Vatican City only came into existence as recently as 1929.

All the earthworms in America would weigh 55 times more than all the American people combined.

In 1999 a three-headed turtle was discovered in Taiwan.

The natives of the Solomon Islands claim to be able to kill trees just by shouting at them.

In Kenya, a person's middle name is based on the time of day at which they were born.

Malta is the nearest Commonwealth country to the UK.

Firsts

In 301, AD Armenia became the first country to make Christianity the state religion.

In 1869 Austria became the first country to use postcards. In 1937 Austria issued the world's first Christmas stamp.

Lego was invented in Demark in 1949. The pieces were originally called Automatic Binding Bricks.

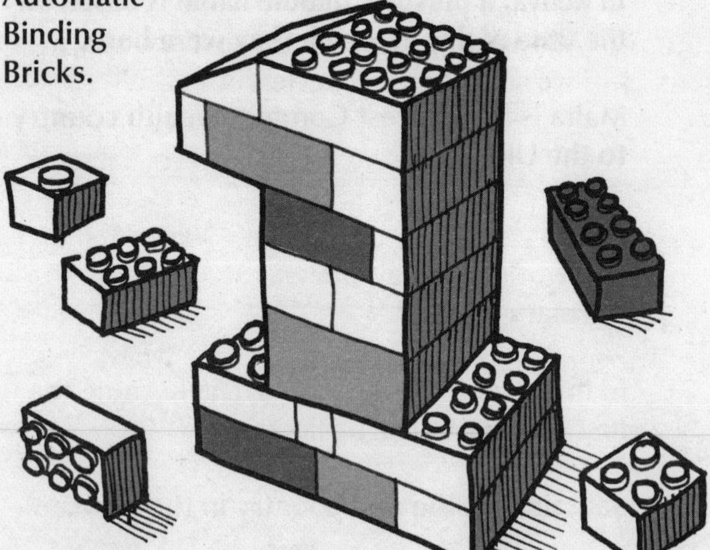

The world's first scheduled passenger air service started in Florida in 1914.

In 1492 Christopher Columbus made his first landfall in the western hemisphere in the Bahamas.

The world's first newspaper was *Relation Aller Fürnemmen und Gedenckwürdigen Historien.* It was published in 1605 in Strasbourg (now part of France but then an imperial free city in the Holy Roman Empire of the German Nation).

In 2004 Bhutan became the first country in the world to ban cigarettes.

The first toothbrush was invented in China in 1498.

The world's first airline was started in Germany in 1909.

In the seventh century BC China became the first country to use banknotes.

Bahrain was the first country in the Arabian

Gulf in which oil was discovered. It was originally discovered in 1902, although it was not fully exploited until the 1930s.

Yoga was first practised in India some 5,000 years ago.

The Egyptians created the first calendar and based it on the flowing of the Nile.

The Egyptians built the first sailing boats.

The first steeplechase for horses was run in County Cork in Ireland in 1752. It derived its name from the fact that riders rode towards a distant landmark like a steeple, jumping over hedges, ditches, banks and walls on the way.

The world's first police force was established in Paris in 1667.

Germany was the first European country to have a McDonald's (in Munich in 1971).

Grenada was the first country to have an Elvis Presley postage stamp.

Coffee was first brewed in Ethiopia. The word coffee comes from Kaffa, the name of a province in southern Ethiopia.

Potatoes were first grown in Peru.

The area we now call Ethiopia is said to be where the very first human beings originated.

In 1986 Japan became the first country in the world to ban lead in petrol.

Mexico was the first country to produce chocolate on a large scale.

Namibia was the first country to make it a political priority to protect the environment.

The Dutch were the first Europeans to discover Australia and New Zealand.

The Norwegians were the first to reach the South Pole.

Oman was the first Arabic country to allow women police officers.

In 1922 Pitcairn Airlines became the first airline to provide sick bags.

You probably know that the Russians were the first to send a man into space (Yuri Gagarin in 1961), but they also sent the first dog into space. In 1957 Laika, a stray, was launched into space on *Sputnik 2*. Alas, she died a few hours after launch from overheating, probably due to a malfunction in the thermal control system. However, the experiment proved that a living passenger could survive being launched into orbit and endure weightlessness.

In 2001 Singapore hosted the first ever World Summit on Toilets.

In 2010 South Africa will become the first African nation to host the FIFA World Cup.

The system of longitude was first discovered by charting the distance between Portsmouth, England, and Bridgetown, Barbados, using the position of the sun in relation to both locations.

Dr Christiaan Barnard performed the world's first heart transplant operation in Cape Town, South Africa, in 1967.

The world's first coffee house was opened in Damascus, Syria, in 1530.

The world's first church was built in Turkey.

In 1930 Uruguay became the first country to host the FIFA World Cup. They also won it.

Fascinating facts!

There's a river in Nicaragua called the Pis Pis River.

There was once an internet rumour that Belgium doesn't exist. That's right, that Belgium – as in the country – doesn't exist. I'll believe it if you do!

The first astronauts to go to the Moon trained in Iceland because the terrain there was reckoned to be similar to the Moon's surface.

Mozambique has all five vowels in it.

If you buy a map of South America in Peru, it'll differ from one sold in Ecuador. This is because there's a big row between the two countries as to who owns the area around the Amazon headwater.

There is no known case of a vegetarian dying from a snake bite in America.

Liechtenstein used to have the world's smallest army. There was one soldier. He served his country faithfully until his death at the age of 95. Then Liechtenstein went from having the world's smallest army to no army at all.

Waikikamukau – pronounced 'Why kick a moo cow?' – is the expression New Zealanders use for a particularly remote rural town. Sadly, there's no actual place with that name!

Girls and women aren't allowed to walk on Mount Athos in Macedonia. In fact, even female animals are not allowed there.

There are people who claim that it's illegal to dress up as Batman in Australia. This is because of an obscure law which prohibits the wearing of dark clothes all over the body for fear that someone will look like a cat burglar. Given that the Batman costume is pretty much all black, some people insist that it is covered by this law and that therefore it must be illegal to dress up as Batman.

In Tibet, some women have special metal instruments they use for picking their noses.

Tobago was Daniel Defoe's inspiration for the island which Robinson Crusoe found himself washed up on.

In the 19th century a French mime artist accidentally got stuck in his imaginary glass box and starved to death. Think about it . . .

There was once a Togolese man with 17 wives and 60 children.

Countries found in England

America, Cambridgeshire
Canada, Hampshire
Egypt, Buckinghamshire
Gibraltar, Buckinghamshire
Greenland, South Yorkshire
Holland, Surrey
Ireland, Bedfordshire
New Zealand, Buckinghamshire
Scotland, Lincolnshire

Foreign places in the UK

California, Buckinghamshire
Dresden, Staffordshire
Jerusalem, Lincolnshire
Maryland, Gwent
Moscow, Scotland
New York, Tyne and Wear
Normandy, Surrey
Pennsylvania, Gloucestershire
Quebec, County Durham
Toronto, County Durham

Way back when

Madagascar became an island 60 million years ago. Before then, it was part of India.

Antigua was first inhabited by the Ciboney (or 'Stone People'), whose settlements date back to at least 2400 BC.

Baghdad was once the centre of the Mesopotamian Empire. The ancient name for most of modern Iraq is Mesopotamia, a Greek word meaning 'between two rivers' – those two rivers being the Euphrates and the Tigris. Most Iraqis still live in this region.

The Faroe Islands were first settled by Irish monks in the sixth century AD.

New Zealand was initially administered as a part of the colony of New South Wales (in Australia). It only became a separate colony in 1840.

Laos was once known as Lan Xang, meaning 'Kingdom of a Million Elephants'.

Martinique was called 'Madinina' (the Island of Flowers) by the Caribs.

Yemen was known to the Romans as 'Arabia Felix' (Happy Arabia) because of the riches it provided for them.

Before the last Ice Age, the islands of Trinidad and Tobago were joined on to the continent of South America.

There's evidence to suggest that some 600,000 years ago, humans inhabited the desolate Sahara of northern Niger.

Moscow's Kremlin was originally a wooden fortress, built in 1156. Over the centuries it was enlarged and is now an enormous complex of government buildings.

The roman names for countries and regions

Latin Name	Current Name
Aegyptus	Egypt
Armorica	Brittany
Belgica	Belgium and the Netherlands
Britannia	Britain
Caledonia	Scotland
Cambria	Wales
Cornubia	Cornwall
Dania	Denmark
Finnia	Finland
Gallia	France
Germania	Germany
Helvetia	Switzerland
Hibernia	Ireland
Hispania	Spain
Islandia	Iceland
Judaea	Israel
Lusitania	Portugal
Norvegia	Norway
Tingitania	Morocco
Tripolitana	Libya

Best

Canadians are the best educated people in the world, with 44 per cent of them going to university.

Berlin is reckoned to be the best city in the world for viewing art.

According to UNICEF, the Netherlands is the best country in the world for children to live in (and Dutch children feel better about their treatment than any other nationality).

The same report found that Sweden is the safest country for children to live in and also the one that provides them with the best quality of life when it comes to material possessions.

Belgium comes top for educational well-being while Italy rates highest for family and friends as far as children are concerned.

The Australian territory of Tasmania has the cleanest air in the inhabited world.

Rwanda has the highest proportion of women in parliament (49 per cent).

The Canadian city of Calgary is the cleanest city in the world.

Environmentally, Finland is the cleanest country in the world.

Andorra has an average life expectancy of over 83 years. As of 2008, this is the highest of any country in the world.

The richest deposit of high quality gold is in Ghana's Ashanti gold fields.

Norway is ranked as having the highest standard of living in the world.

According to statistics, the people of Vanuatu are the happiest people on Earth (they're followed in the happiness stakes by the people of Colombia, Costa Rica, Dominica, Panama, Cuba, Honduras, Guatemala, El Salvador, St Lucia, Vietnam, Bhutan, Western Samoa and Sri Lanka).

Andorra has a zero percent unemployment rate. Everyone's employed.

Switzerland has the best recycling rate in the world. It recycles 52 per cent of all its waste. The Swiss are leaders in recycling everything from cans, glass and plastic bottles, to newspapers, furniture, batteries and clothing.

CANS
PAPER
CUCKOO CLOCKS

Worst

Sumgayit in Azerbaijan is the world's most polluted city, while Baku in Azerbaijan is the world's worst city when it comes to health and sanitation (with Antananarivo, Madagascar, coming second).

Bosnia Herzegovina has the world's worst unemployment rate at over 75 per cent.

Australians are the worst – that's to say biggest – gamblers in the world.

The world's worst volcanic eruption happened on 8 May 1902 at Mount Pelée in Martinique. Nearly 30,000 people were killed.

The world's most active volcano is Sangay in Ecuador erupting once a day on average. It once erupted more than 400 times in a single day.

Canada produces more waste per person than any other country.

Malawi is the poorest country in the world.

The world's most extreme tides occur in the Bay of Fundy in Nova Scotia, Canada, where the water level rises and falls more than 15 metres every six hours. The tide comes in so fast that it can easily overtake a person trying to outrun it.

The UK has the highest proportion of overweight (as opposed to obese) people (over 39 per cent), Mexico is second and Austria third.

The US has the highest proportion (over 30 per cent) of obese people. Mexico comes second and the UK third.

The world's worst flood happened in the Huang He River (also known as the Yellow River) in August 1931. It killed an estimated 3.7 million people.

The world's worst earthquake happened in Tangshan, China, on 28 July, 1976. It killed nearly a quarter of a million people. Tangshan has been called the Brave City of China because of its successful rebuilding efforts.

Indonesia has 146 endangered mammals – more than any other country (followed by India, China and Brazil).

Baghdad is rated the most dangerous city in the world.

Because Australians spend most of their spare time on the beach or participating in outdoor activities, they have the worst (highest) incidence of skin cancer in the world.

On 12 November, 1970, Bangladesh suffered the world's worst cyclone. There were an estimated 200,000 deaths.

The Russians suffered more casualties during the Second World War than the whole world suffered during the First World War.

Biggest

The biggest gold nugget – weighing 235 kilograms – was found in Australia in 1872.

At 6,071 metres high, Guallatiri in Chile is the world's biggest volcano. It last erupted in 1959.

The US has the biggest air force in the world (followed by Russia, China and Israel).

The biggest game of Pass-the-Parcel was played in Singapore in 1998. It involved thousands of people removing over 2,000 layers of wrapping paper from a parcel.

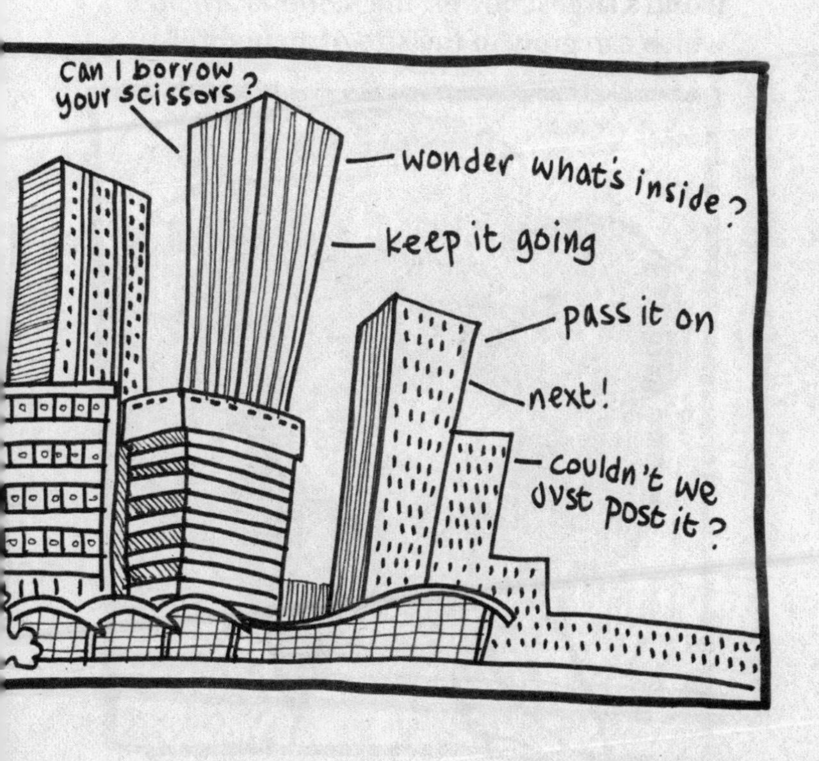

The world's biggest zoo is in Berlin, Germany. The Zoologischer Garten has 1,500 species and more than 14,000 animals.

The world's biggest emerald is on display in Vienna, Austria.

The Indonesian island of Sumatra has the world's largest flower: the *Rafflesia arnoldii,* which can grow to the size of an umbrella.

Thailand is home to the biggest crocodile farm in the world.

The ancient Borobudur Temple in central Java is the world's biggest Buddhist monument. It was built between 778 and 850 by more than 10,000 labourers and carvers, using only hand tools.

The world's biggest baby – Samuel Timmerman – was born in Belgium in 2006. He weighed 5.4 kilograms.

Istana Nurul Iman is the biggest residence in the world. It's the palace of the Sultan of Brunei and has 1,788 rooms and 257 bathrooms.

The biggest lizards in the world are the Varanus (or monitor) lizards of New Guinea which grow up to five metres long.

The world's biggest shopping mall is in Dubai.

Japan boasts the biggest statue in the world: the Bronze statue of Buddha in Tokyo, which

is 120 metres high and 35 metres wide.

The biggest McDonald's in the world is in Beijing.

Australia's aptly named Great Barrier Reef is the biggest in the World – followed by Belize's barrier reef which runs along its entire coastline.

The world's biggest ruby was found in Thailand in 1999.

Bolivia is the biggest exporter of Brazil nuts (and, no, Brazil isn't the biggest exporter of Bolivian nuts).

The biggest religious building in the world is a Hindu temple in Cambodia.

The world's biggest sports dome is in Qatar.

The pla buek is the biggest freshwater fish in the world, measuring up to three metres in length and weighing about 300 kilograms. It can be found in the Mekong River in Asia.

The world's two biggest cuckoo clocks are both located in Germany. One of the cuckoos is nearly five metres high and weighs 150 kilograms.

Berlin boasts Europe's biggest train station.

The biggest tree trunk in the world – a Montezuma cypress measuring 14 metres in diameter – can be found in the grounds of Santa María del Tule in Oaxaca, Mexico.

The tallest tree – and the largest living thing on Earth (by volume) – is the General Sherman tree in Sequoia National Park. It's 84 metres tall with a trunk measuring 11 metres in diameter at its widest point.

Relative values

The whole of the Caribbean island of Anguilla is only the size of Birmingham.

Vatican City and Monaco are both smaller than New York's Central Park.

Monaco's national orchestra is bigger than its army.

Every year, there are more births in India than there are people in Australia.

China has more English speakers than the United Kingdom.

The island of Antigua is twinned with the island of Barbuda for administrative purposes (rather like Trinidad and Tobago). As well as the main two islands of Antigua and Barbuda, there are also several much smaller islands, including Prickly Pear Island, which has just 12 residents – all of whom were born on the island. Incidentally, its total population wouldn't be enough for a rugby team but they

could start their own football team – with a substitute!

There's a shopping mall in Washington, DC that's bigger than the Vatican.

There are more Barbie dolls in Italy than there are Canadians in Canada.

Norway's total coastline is longer than that of the United States although its landmass is 27 times smaller.

Luxembourg is smaller than Rhode Island, the smallest state in America.

There are enough horses in the world for every man, woman and child in the UK to own one.

There are six times as many sheep in the UK as there are in the US.

Bangladesh's population is about half that of the United States, yet it's only about the size of New York.

Australia has a third of the population of Britain, yet is 31 times bigger.

There are more species of tree in Singapore's Bukit Timah Nature Reserve than there are in the whole of North America. Even more extraordinary when you consider that the whole of the nature reserve fits into less than two square kilometres!

Afghanistan, Botswana, Burma and the Ukraine are all smaller than the state of Texas.

Antarctica is nearly one and half times as big as the US.

If Texas were a country its Gross National Product (GNP) would be the fifth largest of any country on earth.

There are 10 times more horses than people in Mongolia.

Brazil, Botswana, Argentina, Australia and New Zealand all have more cattle than people.

There are more penguins than people in the Falkland Islands.

There are more kangaroos than people in Australia.

Countries with more than 100 languages

Papua New Guinea: 820 languages
Indonesia: 742
Nigeria: 516
India: 427
US: 311
Mexico: 297
Cameroon: 280
Australia: 275
China: 241
Democratic Republic of the Congo: 216
Brazil: 200
Philippines: 180
Malaysia: 147
Canada: 145
Sudan: 134
Chad: 133
Russia: 129
Tanzania: 128
Nepal: 125
Vanuatu: 115
Myanmar: 113
Vietnam: 104

Onlys

There's a rare species of lizard – the Antiguan ground lizard – which can only be found on the island of Antigua. Clue's in the name, you'd have thought.

Albania was the only European country occupied by the Axis powers (that's Germany and Italy) that had a larger Jewish population after the Second World War. Only one Jewish family was deported and killed during the Nazi occupation of Albania. Not only did the Albanians protect their own Jews, but they provided refuge for Jews from neighbouring countries.

Tonga's the only island nation in the Pacific Ocean that's still a monarchy. Tonga is also the only country ever to have issued a banana-shaped stamp.

South Africa has only one overseas possession: the Prince Edward Islands, a small sub-antarctic archipelago.

The Dutch town of Abcoude is the only town in the world whose name begins with ABC.

Barbados was the only foreign land that George Washington, the US's first president, ever visited.

Thimphu in Bhutan is the only capital city in Asia that doesn't have traffic lights. They use traffic policemen instead.

Brazil is the only country to have played in every football World Cup. They've won it five times (most recently in 2002), which is more than any other nation.

Colombia is the only country in South America that has a coast on both the Pacific and Atlantic Oceans.

A small number of Weddell seals live on South Georgia, the Falkland Islands, the only group outside the Antarctic continent.

Nicaragua's Volcán Masaya is a national park that's home to Crater Santiago, the only Central American crater where you can see molten lava. Despite the fact that the crater emits strong sulphuric fumes, green parakeets nest and roost on its walls.

Qatar is the only country that starts with a Q.

Iraq is the only country that ends with a Q.

Of the Seven Wonders of the World, only the Pyramids of Giza in Egypt are still in existence. Incidentally, the inside of the Great Pyramid at Giza is always 20 degrees Celsius.

In 2008 New Zealand was the only country in the world in which all the highest offices in the land were occupied by women – the Sovereign Queen Elizabeth II of New Zealand, Governor-General Dame Silvia Cartwright, Prime Minister Helen Clark, Speaker of the New Zealand House of Representatives Margaret Wilson and Chief Justice Dame Sian Elias. That finished in November 2008, when Helen Clark lost the general election and was replaced as Prime

Minister by John Key.

Panama is the only place in the world where you can see the sun rise in the Pacific and set in the Atlantic.

Saudi Arabia, named after King Saud, is the only country named after its ruling family.

Summer is the only season in the Seychelles.

Tonga's only golf course has just 15 holes. And there's no penalty if a monkey steals your ball.

The most visited countries in the world

France
Spain
China (inc. Hong Kong)
US
Italy
UK
Austria
Mexico
Germany
Canada
Hungary
Greece
Poland
Turkey
Portugal
Malaysia
Thailand
The Netherlands
Russia
Sweden

Countries at risk from flooding

Bangladesh is one of the countries most at risk from rising sea levels. Much of the country would be underwater if sea levels rose.

The Îles Éparses (literally 'scattered islands') are off the coast of Madagascar in the Indian Ocean and have no permanent population – which is just as well as they're only two metres above sea level. Administered by the French, the islands are mainly used for meteorological purposes (e.g. cyclone warnings).

80 per cent of the 1,200 islands that make up the Maldives are no more than a metre above sea level. Within 100 years the Maldives could become uninhabitable.

Half of the Netherlands is less than a metre above sea level. A lot of it is below sea level. Disastrous flooding in 1953 killed 1,800 people and destroyed more than 70,000

homes. This led to a huge flood-control project. Flooding struck again in 1995. Rising rivers forced 240,000 people to move out of their homes until the water receded.

Number one (consumption)

THEY CONSUME THE MOST . . .
(* = per person)

Rice: China (32 per cent of the world's rice)
Fresh pork: China
Turkey-based products*: Israel
Bread*: Poland
Coca-Cola*: Mexico
Eggs*: Japan
Toilet paper: US (in just one day, Americans use enough toilet paper to wrap around the world nine times. If all the toilet paper they used were on one giant roll, they would be unrolling it at the rate of 7,600 kilometres per hour)
Olive oil*: Greece
Honey*: Greece
Chocolate*: Belgium
Alcohol*: Luxembourg
Olives*: Syria
Meat: US
Meat*: US (an average 120 kilograms per person per year)

Pasta*: Italy (an average 27 kilograms per person per year)
Sugar*: Singapore
Tomato ketchup: US
Tomato ketchup*: Sweden
Beer*: Czech Republic (an average 160 litres per person per year)
Sweets*: The Netherlands
Wine*: France (an average 60 litres per person per year)
Bottled water*: France
Tea*: Ireland (an average 3.2 kilograms per person per year)
Baked beans*: UK
Crisps*: UK
Coffee*: Finland (an average 4.5 cups of coffee per person a day)
Butter*: France (an average 8.5 kilograms per person per year)
Cheese*: France (an average 23 kilograms per person per year)
Potatoes*: Ireland (an average 170 kilograms per person per year)
Milk*: Ireland (an average 165 litres per person per year)

Number one (production)

THEY PRODUCE THE MOST . . .

Apples: China
Apricots: Turkey
Attar of roses: Bulgaria (about 70 per cent of the world's attar of roses – an ingredient in the most expensive perfumes: 2000 petals are needed to make a single gram of attar of roses)
Bananas: Ecuador
Beer: US
Butter: India
Camels: Somalia
Cars: Japan
Cheese: US
Coal: China
Cocoa: Ivory Coast (produces almost half the world's cocoa) 
Coffee: Brazil
Cotton: China
Diamonds: Australia
Flowers: The Netherlands
Goats: China

Gold: South Africa (produces two-thirds of the world's gold)
Ice-hockey pucks: Slovakia
Iron: China
Lead: Australia
Lemons: Mexico
Oil: Saudi Arabia (the discovery of oil in Saudi Arabia in the 1930s improved the country's economy tremendously: today, Saudi Arabia's oil reserves are estimated at 260 billion barrels)

Rice: China
Salmon (farmed): Norway
Salt: US
Sheep: China
Silver: Mexico
Tea: India
Tomatoes: China
Toys: China

Vanilla: Madagascar (supplies over half the world's vanilla. Over three-quarters of the vanilla beans used to make vanilla ice cream are grown here. The Madagascan economy suffered briefly in the 1980s when Coca-Cola stopped using as much vanilla in drinks)
Wheat: China
Wine: France
Wood: US
Wool: Australia
Yams: Nigeria
Zinc: China

Population

With over 1.3 billion people, China is the most populated country on earth. At present, China constitutes one-sixth of the world's population. If the people of China walked past you in single file, the line would never end because of the rate of reproduction.

Niger has the highest birth rate in the world (nearly 50 births per year for every 1,000 people).

Germany has the lowest birth rate in the world (about eight births per year for every 1,000 people).

There are 13 countries where people born today can expect to live to over the age of 80. They are (from the highest):

Andorra (83.5), Macau (82.2), Japan (82.02), San Marino (81.80) Singapore (81.80), Hong Kong (81.68), Sweden (81.68), Australia and Switzerland (80.62), France (80.59), Iceland (80.43), Canada (80.34) and the Cayman Islands (80.20). Britain comes in at 78.7 – lower than New Zealand, but higher than the US.

There are 27 countries in the world where people born today can expect to live to the age of only 50, and shockingly all are in Africa, apart from Afghanistan. And there are five countries in Africa where people born today have a life expectancy of less than 40 – that's under half the age of those born in the richest, healthiest countries of the world. They are Lesotho, Zimbabwe, Zambia, Angola and Swaziland.

In many African countries, half the population is under the age of 15.

Fewer than three per cent of Ugandans are above the age of 65. Contrast this with Monaco, which has the oldest population in the world, with nearly a quarter of its residents aged over 65. Britain's not that far behind, with 16 per cent of its people over 65.

As well as being one of the smallest countries in the world, Monaco is also the most densely populated. It has nearly 33,000 people living in less than two square kilometres, which means over 20,000 people live in each square kilometre, compared with 246 people per square kilometre in Britain.

Mongolia is the most deserted country on the planet, with fewer than two people (1.7) living in each square kilometre. That's excluding the Falkland Islands (0.25) and Greenland (0.026).

With 131 million people, Nigeria has the highest population of any African country.

With 35 million people living there, Tokyo is the most populated city in the world. A hundred years ago, that honour belonged to London, with seven million residents. Nowadays, with a similar number of people, it's not even in the top 10. However, it's still the largest city in the UK – followed by Birmingham.

The US is 49.1 per cent male and 50.9 per cent female which means that there are five million more women than men. Only the states of Nevada and Alaska have more men than women.

The world's population grows by about 250,000 people every day (that figure takes into account all the people who die). If that rate of increase continues, the world's population will double to more than 12 billion in fewer than 50 years.

There are as many Italians living in other countries as there are living in Italy.

El Salvador is the most densely populated country in Central America, with an average of over 290 persons per square kilometre.

The Dutch got round their over-population problem by reclaiming land from the sea. This meant draining the sea water to create new land.

Clothes

Some East African women wear a *kanga*, a brightly coloured cloth wrapped around their bodies and heads. Kangas were named after a spotted bird because many of the original designs featured white dots on a dark cloth.

In Cambodia, classical dance troupes wear tight-fitting costumes which are so tight they have to be sewn onto the dancers just before the performance.

Vietnamese farmers wear conical hats to keep the sun off their heads. Some of the hats have traditional poems inscribed inside the brim.

Whereas in Britain the colour of mourning is black, in Turkey it is violet. In China the colour of mourning is white.

Traditionally, Chinese brides wear red – not white.

The Aymara women of Bolivia, Peru and Chile wear their bowler hats tipped to one side if they're single, but in the middle of their heads if they are married.

Congolese women wear a *pagne*, a long dress made of a five-metre length of fabric.

Language (1)

There's a Mexican language named Zoque which withered away until there were just two people in the whole world who spoke it. The trouble was that those two men – both in their 70s – hated each other and so refused to speak.

There are a staggering 7,000 languages in the world. Mandarin is the most widely spoken, followed by English and then Spanish. Even though Russia is one of the biggest countries in the world, there are only 145 million people who have Russian as their first language.

The German language combines words to make composite words. So, for example, the single German word for 'favourite break-time sandwich' is 'Lieblingspausenbrot'.

Romanian words sound similar to some Italian, French or Spanish words because Romania is the only country in Eastern Europe where people speak a language of Latin origin.

There are 32 letters in the Polish alphabet, including three variations of the letter Z.

The Lao language has no words ending in 's'. So *they* call their country Lao instead of Laos.

The Lozi language of Zambia has at least 40 words meaning 'woman'. Each describes a woman at a particular stage in life. For example, an unmarried woman, a newlywed, just arrived in her husband's village, a widow, etc.

In the Malay language, plurals are formed by repeating the word. For example, the Malay word for man is 'laki' so the Malay word for men is 'laki-laki'.

The Japanese equivalent of *woof-woof* (a dog's bark) is *wan-wan*.

There are African languages that have fewer than 100 people speaking (each of) them. The speakers tend to live in closed communities cut off from most of the rest of the world. As the world discovers them and they discover the world, they're obliged to learn a more popular language and so their native tongue dies.

Nearly half of all Germans are fluent in English, but only three per cent speak French fluently.

In Bulgaria, shaking the head from side to side means yes and nodding up and down means no.

Because of the mix of cultures and languages, many Mauritians are multilingual. A person might speak Bhojpuri at home, French to a supervisor at work,

English to a government official and Creole to friends.

English teenagers learn the fewest languages in Europe. On average, they learn 0.6 of a language (I know it sounds preposterous but it *is* an average!), as against the 1.4 languages that the average teenager learns in other EU countries.

The Cambodian alphabet has 72 letters.

Many people in the Syrian town of Ma'loula still speak Aramaic, the language spoken by Jesus of Nazareth.

In Hungary, the word 'szia' can be used to mean either 'hello' or 'goodbye' when it's said to an individual. This probably explains why Hungarians whose English isn't that good will sometimes say 'hello' instead of 'goodbye' when they're leaving.

That's surprising! (2)

Although the Jewish religion forbids the eating of pork, there are 30 pig farms in Israel.

Hong Kong is very hilly and there are outdoor escalators in the Central District of Hong Kong Island (I've ridden on them to get from one part of the city to another and it feels very strange indeed).

A possum once travelled 12,000 miles from New Zealand to Felixstowe in the UK in a crate of onions.

The Red Sea is a unique body of water. Because no rivers flow into it, the sea has a higher salt content than most oceans.

Sweden's capital city Stockholm is built on 14 islands. Water covers one-third of the city's area, and it's so clean that you can catch salmon right in the city centre.

After the Norman invasion of 1066, French was the official language of England for 300 years.

The temple at Ġgantija, Malta, was built more than 5,500 years ago – even earlier than the Egyptian pyramids.

The Minangkabau people of West Sumatra in Indonesia have a matrilineal society women own all the property and only daughters can inherit. In traditional Minangkabau families, the men live with their mothers and visit their wives.

Two of the Seven Wonders of the World were in Turkey.

People from Niger are Nigerian – as are people from Nigeria. This must get confusing.

The Bahamas are made up of 700 islands but only around 30 of them are inhabited. There are also a staggering 2,000 much smaller islands (some of which are little more than very large rocks).

There is a house in Margate, New Jersey US, that is made in the shape of an elephant. A home in Norman, Oklahoma, is shaped like a chicken.

Temperature has such an effect on the height of the Eiffel Tower that, in practice, it has no fixed height.

Paper bags are outlawed in grocery stores in Afghanistan. They believe paper is sacred.

Magnetic north is actually 1,000 miles away from the North Pole ('true north'). The North Pole is technically located at 90° N (& any longitude), but magnetic north is currently at 73° N, 100° W.

Less than one per cent of Caribbean Islands are inhabited.

Every year, Mexico sinks about 25 centimetres.

Following the release of the film *Borat: Cultural Learnings of America for Make Benefit Glorious Nation of Kazakhstan,* the number of tourists visiting the country increased significantly.

Volcanoes aren't always bad. In Guatemala, the town of Fuentes Georginas has hot baths and steam rooms heated entirely by volcanic heat.

At a recent Mongolian election, something

wonderful happened: there was a tie between the two rival parties so they just shared the Premiership, with one party leading for the first two years before handing over to the other party for the next two years. That's a nice change from the Genghis Khan days . . .

Switzerland stayed out of the two world wars despite being (geographically) in the middle of all the fighting. Yet, in spite of their constant neutrality, the Swiss Constitution states that 'every Swiss male is obliged to do military service'.

Cannibalism went on in Fiji well into the 19th century.

In Japan they sell square watermelons.

Genuine dishes from around the world

Stir-fried Dog (China)
Rabbit Excrement (Red Indians of Lake Superior used this as a flavouring in red wine)
Minced Giant Bullfrog Savoury Sandwich Spread (US)
Deep-fried Horsemeat (Switzerland)
Mixed Organ (spleen, pancreas, aorta, etc.)
Beef Stew (Austria)
Caterpillar Lava Of The Large Pandora Moth (Pai-utes Indians of Oregon)
Roast Wallaby (Australia)
Calf's Head with Brain Fritters (19th-century US)
Steamed Cat and Chicken (China)
Burgoo (squirrel, rabbit, pigeon, wild duck and/or chicken, vegetables stew) (U.S. Appalachian)
Bandicoote Stewed In Milk (Australia - early 20th century)

Pork Intestines With Fish Cake and Liver (China)
Pea Soup with Pigs' Ears (Germany)
Tripe Soup (Czech Republic)
White Ant Pie (Zanzibar)
Pigs' Tails (France)
Smoked Dog (Philippines)
Crisp Roasted Termites (Swaziland)
Dragon, Phoenix and Tiger Soup (consists of snake, chicken and cat) (China)
Roasted Palmworms with Orange Juice (French West Indies)
Baked Opossum (US)
Calf's foot Stew (Philippines)
Stewed Veal Shins (Italy)
Pot-roasted Cow's Udder (France)
Bear's Paws Dalmatian Style (Croatia and Dalmatia)
Dog Ham (China)
Pigs' Ears (Germany)
Baked Elephant Paws (Africa – 19th century)
Broiled Puppy (Hawaii)
Boiled Locusts (Vietnam)

Mosts

Anguilla has the highest marriage rate in the world.

Argentinians listen to the radio more than any other people (over 20 hours per week on average).

Japan is the most expensive country in the world to live in.

New Zealand boasts the steepest street in the world: Baldwin Street in Dunedin, which has an incline of 38 per cent.

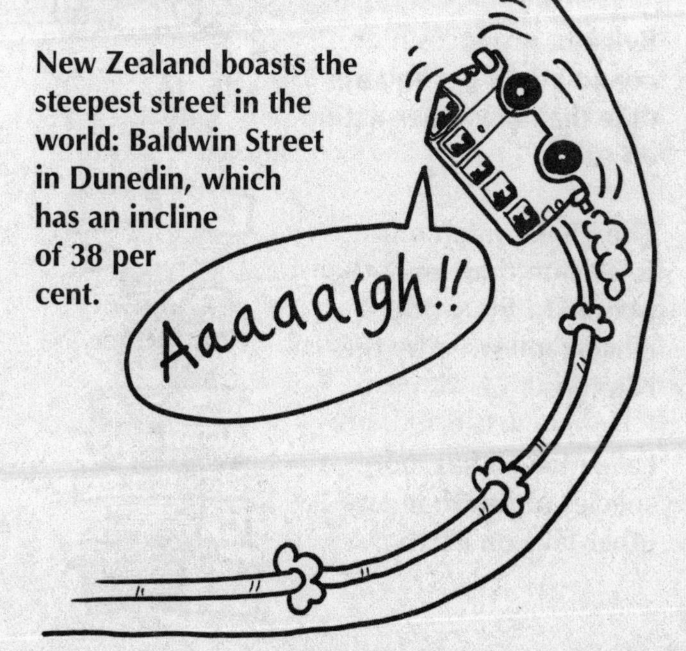

Australia has more cattle than people. It also has more sheep than people. In fact, it has the highest number of sheep per person of any country.

Sweden has the highest proportion of people over 65.

Icelanders read more books than any other people in the world.

Belgium produces more comic books per square mile than any other nation on earth.

The Thais watch more television than any other people in the world – closely followed by the Filipinos.

Lake Malawi has more species of fish than any other lake on Earth.

With more than 200 million paying visitors since it was built, the Eiffel Tower in Paris is the world's most popular monument (of all those that charge admission).

The Eiffel Tower took 300 workers two years to build. It has over 18,000 iron pieces. It's repainted completely every five years. It takes about 15 months and 60 tons of paint to cover the tower, with 25 painters using only brushes. The paint colour is called *Eiffel Tower brown* and it comes in three shades, with the lightest shade at the top and the darkest at the bottom.

The Taiwanese spend more time browsing the internet than any other people.

Colombia has more plant and animal species per square kilometre than any other country in the world.

Israel has the highest percentage of home computers per person.

France is the most visited country, with over 71.4 million visitors each year.

The United States consumes 25 per cent of the world's energy.

Indonesia is the country with the most volcanoes.

China consumes 33 per cent of the world's iron ore and 40 per cent of the concrete.

China uses 45 billion chopsticks per year.

Americans eat the most meat, drink the most fizzy drinks and buy the most toys.

Venezuela has won the Miss World and Miss Universe contests the greatest number of times.

The UK has more public-lending libraries than any other country in the world.

Frankfurt International Airport in Germany sends planes to more international destinations than any other airport.

Reykjavik is the most northern capital city in the world.

Shanghai in China is the busiest port in the world; Rotterdam is the second busiest in the world and the busiest in Europe.

Alaska is the most northern and western state of all the United States. And because it crosses the International Date Line, you could say that it's also the most eastern state. On top of all that, it's bigger than the 21 smallest states combined.

On average, the Dutch are the tallest people in Europe.

Proportionally, Denmark has won the highest number of Nobel Prizes.

Singapore and Luxembourg share the distinction of being the most 'Googled' countries in the world.

Liberia has the biggest annual population growth (+6.2% / year).

The UK publishes more books than any other country in the world.

Ecuador has over 2,000 endangered species – more than any other country.

China has the highest number of border countries in the world (15).

Canada has the world's highest proportion of left-handed people.

Wellington, the capital of New Zealand, has more restaurants per person than any other capital city. It also has the cleanest air of any capital city in the developed world.

Hong Kong has the world's highest ratio of restaurants to people: one for every 20.

The most famous Persian carpet in the world is the 16th-century Ardabil carpet, which is on display at the Victoria & Albert Museum in London. Measuring 10.7 by 5.3 metres, it has a million knots (by comparison, a really top carpet today will have 500).

There are more public holidays in Sri Lanka than there are in any other country.

Botswana has the largest population of elephants in the world.

Brazil borders 10 countries: that's every country in South America except Chile and Ecuador.

Rome has more homeless cats per square kilometre than any other city in the world.

Asia is the continent with the most people, the most land and the highest mountains.

China has the largest number of hospitals in the world, with over 67,807.

Estonia has more meteor craters per square kilometre than any other country.

India is the world's largest consumer of gold.

France has more pets per person than anywhere else.

Canada has more lakes than the rest of the world combined, and one-third of all the fresh water on Earth is in Canada.

The furthest point from any ocean is in China.

The country with the highest density of trees in the world is Suriname, the smallest country in South America: it is 94 per cent forested.

America, which is only 33per cent forested, produces the world's most timber; it also has all the world's tallest trees, mostly in its national parks.

The country with the most dentists per person is America.

San Marino has the most nurses and the most doctors per person.

Finland is the country with the highest proportion of natural blondes.

The Seychelles has the world's biggest population of giant turtles.

Japan and Barbados have the highest number of centenarians – that's people who live to the age of 100 or beyond – in the world.

The UK has the highest number of botanical gardens and zoos in the world (relative to its size).

Singapore holds the record for the greatest number of people involved in a line dance: 11,967.

The Amazon River carries more water than any other river and is the source of 20 per cent of the Earth's water.

20 per cent of the world's oxygen is produced by the Amazon rainforest (due to the masses of plants photosynthesizing).

Sport & leisure

As anyone who's read the moving novel *The Kite Runner* – or seen the film – will know, a game popular with young people in Afghanistan is 'kite fighting'. Kite strings are covered with a mixture of flour and powdered glass and participants try to cut through the strings of other kites.

Wrestling is also popular in Afghanistan. The object of the game is to pin your opponent to the ground without touching his legs.

Before we leave Afghanistan, let me draw your attention to a game they play in the countryside which requires players to hold their left feet in their right hands and hop around trying to knock each other over.

The Bangkok Golf Club is floodlit so it is possible to play 24 hours a day.

There's an official snowball-throwing contest in Sweden.

Dutch farmers learned a kind of pole-vaulting so they could cross drainage ditches and get from field to field. This became a sport called *polsstokspringen*. But that's not the only game for which the Dutch might be responsible. Some historians say they invented baseball. Art from the 1600s shows children playing something very similar to the modern game. So it's possible that Dutch settlers may have brought baseball to North America.

Yukigassen (a Japanese word which means 'snow battle') is a sport that started in Japan but is also popular in Finland, Norway and Australia. Two teams with seven players each chuck snow at each other on a court with specific measurements. Sounds fun!

In Albania, a traditional sport for women is

competitive mountain climbing against the clock.

The official sport for the state of Maryland, US, is jousting.

Brazil has more professional soccer teams than any other country in the world.

Bandy is a winter sport that's a cross between football and ice hockey.

Canada's official national sports are ice hockey (in the winter) and lacrosse (in the summer).

One sport played at festivals in the desert regions of Algeria requires riders to shoot at a target while galloping at full speed, before bringing their horses to a complete stop. Another tradition during festivals is camel dancing, in which riders urge their camels to move to the beat of traditional music.

Chess is so popular in Armenia it's practically the national sport!

Only four countries participated in all of the first three football World Cups: Romania, Brazil, France and Belgium.

Every August, an unusual race is held in the Australian Outback near Alice Springs. It's called the Henley-on-Todd boat race but it's a boat race with a difference. Because water is rarely found in the Todd River, competitors must carry their boats along the dry riverbed.

The game Monopoly was very popular in Cuba but, because it was considered a game of capitalism, Fidel Castro ordered that all boards be destroyed.

Sri Lankan kites can be enormous, up to nine square metres. People have to use smaller kites just to get the bigger ones into the air and they require up to 90 metres of string to fly.

Cycling is the number one sport in Belgium,

which has hundreds of kilometres of cycling paths.

The only European countries where football isn't the most popular spectator sport are Ireland (Gaelic football), Finland (ice hockey), and Latvia and Estonia (both basketball).

Buzkashi, which literally means 'goat grabbing', is a popular sport in Central Asian countries. It dates as far back as the 13th century. Here's how it works: the mounted players have to ride around a pit that contains a dead calf. They all try to pick it up and once one of them has succeeded, he has to drop it into the 'Circle of Justice' – the goal. The other riders have to try and stop him. The only loser is the calf . . .

Finland is home to a sport named *pesäpallo*. It's a summer sport played in rural areas and small towns and is very similar to baseball.

A British man named Dave Cornthwaite took up skateboarding in March 2005: 14 months later he became the first person to skate the 1,450-kilometre journey from John o'Groats

to Land's End.

Kabaddi is the national sport of Bangladesh but it's also popular throughout South-East Asia. In Kabaddi, two teams occupy opposite halves of a field and take turns to send a 'raider' into the other half to win points by tagging or wrestling members of the opposing team. The raider then tries to return to his own half, holding his breath during the whole raid. Each team consists of 12 players (seven on the field with five in reserve). It's an exhausting game so it consists of just two 20-minute halves. The British army now plays the game as part of its training.

Bizarre place names

BIZARRE PLACE NAMES IN THE US

Accident, Maryland
Arsenic Tubs, New Mexico
Big Ugly Wilderness Area, West Virginia
Bottom, North Carolina
Chargoggagoggmanchauggagoggchaubun-agungamaugg, Massachusetts
Cheesequake, New Jersey
Coolville, Ohio
Cut and Shoot, Texas
Cut Off, Louisiana
Dismal, Tennessee
Double Trouble, New Jersey
Dry Prong, Louisiana
Duel, Michigan
Elephant Butte, New Mexico
Foggy Bottom, Washington, DC
Forks of Salmon, California
Friendly, Maryland
Gas, Kansas
Gnaw Bone, Indiana
Grandmother Gap, North Carolina
Gross, Nebraska

Gun Barrel City, Texas
Hot Coffee, Mississippi
Jackpot, Nevada
Jobsville, New Jersey
Manunka Chunk, New Jersey
Monkey's Eyebrow, Kentucky
Neck City, Missouri
No Name, Colorado
Pie Town, New Mexico
Santa Claus, Indiana
Welcome, South Carolina
Winter, Wisconsin
Wynot, Nebraska
Zap, North Dakota
Zzyzx, California

BIZARRE PLACE NAMES IN THE UK

Beer, Devon
Bunny, Nottinghamshire
Catbrain Hill, Gloucestershire
Christmaspie, Surrey
Cold Christmas, Hertfordshire
Foul Mile, West Sussex
Foulness, Essex
Frisby on the Wreake, Leicestershire

Great Fryup, North Yorkshire
Heart's Delight, Kent
Lickey End, Worcestershire
Mumbles, Swansea
New Delight, West Yorkshire
No Place, County Durham
Nose's Point, Durham
Pease Pottage, West Sussex
Pratt's Bottom, Kent
Puddletown, Dorset
Queen Camel, Somerset
Rest and Be Thankful, Argyll and Bute
Sandwich, Kent
Toot Baldon, Oxfordshire
Twenty, Lincolnshire
Ugley, Essex
Westward Ho!, Devon (the punctuation mark is part of the name)
Woon Gumpus Common, Cornwall

BIZARRE PLACE NAMES IN THE REST OF THE WORLD

Apples, Switzerland
Bad Kissingen, Germany
Bitey Bitey, Pitcairn Island
Bong Bong, Australia

Bulls, New Zealand
Burrumbuttock, Australia
Chinaman's Knob, Australia
Coca Cola, Panama
Egg, Austria
Frenchman's Butte, Canada
Halfway House, Canada
Hells Gate Roadhouse, Australia
Hotazel, South Africa (pronounced 'hot-as-hell')
Howlong, Australia
John Catch a Cow, Pitcairn Island
Kissing, Germany
Maidslain, Canada
Middelfart, Denmark
Moose Jaw, Canada
Mutters, Austria
No Guts Captain, Pitcairn Island
Ogre, Latvia
Police, Poland
Puke, Albania
Punkeydoodles Corners, Canada
Rum, Austria
Saint-Louis-du-Ha! Ha!, Canada
Salmon Arm, Canada
Terry Hie Hie, Australia
Tubbercurry, Ireland

U, Panama
Wagga Wagga, Australia
Where Freddy Fall, Pitcairn Island
Where Reynolds Cut The Firewood, Pitcairn Island
Worms, Germany

BIZARRELY NAMED LAKES, RIVERS, HILLS & MOUNTAINS, etc.

Awe, a loch in Scotland
Bang Bang Jump Up, a rock formation in Australia
Blow Me Down, a provincial park in Canada
Cadibarrawirracanna, a lake in Australia
Darling, a river in Australia
Grandfather, a mountain in the US
Great Slave, a lake in Canada
Hungry Law, a peak on the border of England and Scotland
Hopeless, a mountain in Australia
Piddle, a river in Dorset
Pis Pis, a river in Nicaragua
Possum Kingdom, a lake in the US

Communications

More than half the people in the world have never made or received a telephone call.

The Russian space station, Mir, had a post office so that cosmonauts could send messages to Earth.

In Britain, it's reckoned that every time broadband grows by 1 per cent, newspaper sales decline by 0.2 per cent.

Algeria has the biggest percentage of homes with satellite TV.

In Britain, eight is the average age at which a child gets a mobile phone.

Filipinos send more text messages per day than all the people in Europe and the USA combined.

Countries where people own the most mobile phones: China, 461 million; US, 233 million; India, 166 million; Russia 133 million; Japan (where most of them are made), 101 million; Brazil, 99 million; Germany, 84 million; Italy, 71 million; UK, 69 million. That's still more than one for each person in the UK, which is slightly odd if you consider that lots of people don't have a mobile phone. Obviously, there are some people out there who have several!

Qatar is the cheapest place in the world to buy mobile phones. However, the phone number 666 6666 fetched £1.4 million at an auction (the number six is considered lucky in Qatar).

And, finally, the few countries of the world where a mobile phone is useless due to total lack of reception are: The Falkland Islands, Tokelau, Tuvalu, Wallis and Futuna in the South Pacific, and the western Sahara.

History (1)

During Celtic times, warriors on the front line of the Irish army would go into battle completely naked.

For a long time Ethiopia was ruled by women. That explains why Alexander the Great stopped at the border: he didn't want to be defeated by women.

The Moroccan city of Tangiers was once governed by eight European countries at the same time.

French women didn't get the vote until 1945. Before 1964, when the Matrimonial Act was passed, French women couldn't open a bank account or start a business without their husband's permission.

The US bought Alaska from the Russians in 1867 for two cents per acre ($7.2 million in total).

Although modern images and descriptions of India often show poverty, India was one of the richest countries in the world till the British arrived.

In the 15th century Prince Henry the Navigator set up a school at Sagres, Portugal, where he gathered the finest map-makers, astronomers and navigators of the age. Ferdinand Magellan and Vasco da Gama were among the great explorers who studied there. Their influence survives to the present day: nine countries have Portuguese as their official language.

Austria lost five wars in the 19th and 20th centuries.

In the Middle Ages Slovakia was a centre for gold and silver mining. In the 15th century it produced 40 per cent of the world's gold and 30 per cent of its silver.

The Great Wall of China, all 6,000 kilometres of it, was built in the third century BC to protect China's heartland from invasion.

Between 1839 and 1855 Nicaragua had 396 different rulers.

The famous English explorer Sir Walter Raleigh visited Guyana in 1595 because he believed that it was the site of El Dorado, the legendary city where even the streets were made of gold. His book describing Guyana's riches inspired many Europeans to seek their fortunes there.

Religion

The biggest religion in the world is Christianity, followed by Islam and then Hinduism. Among Christians, Roman Catholics outnumber Protestants.

Armenians celebrate Christmas on 19 January.

The Kogi people of Colombia consider the Sierra Nevada Mountains to be sacred because they believe they are the centre of the universe.

Being a Buddhist monk is a popular rite of passage for young Cambodian men.

Jerusalem was where the world's three biggest religions all started out.

At least nine out ten Nigerians attend church regularly.

In Russia, Christmas is celebrated on 25 December and 7 January.

There's a special mosque in Timbuktu in Mali which has a door that's never been opened. It is believed that opening that door will herald the arrival of the end of the world.

In the town of Atlántida, just outside Montevideo in Uruguay, is a remarkable church built in 1959. Its brick walls curve in and out like the waves of the sea.

There are over 200,000 monks in Thailand.

Churches in Malta show two different times to confuse the devil.

The glue on Israeli postage stamps is certified kosher.

Moscow's famous St Basil's Cathedral was commissioned by Ivan the Terrible and built on the edge of Red Square between 1555 and 1561. Legend has it that on completion of the church the tsar ordered the architect to be blinded to prevent him from ever creating anything to rival its beauty.

80 per cent of people in the Czech Republic

say they don't believe in God.

There's a church in the Czech Republic that has a chandelier made out of human bones.

Slovakia has a rich architectural heritage of wooden churches. The wooden church in Paludza seats 3,600. Not a single nail was used in its construction.

Dominica has the highest percentage of Christians of any country in the world (99.6%).

Most Muslims live in Pakistan, most Hindus in India and most Buddhists in China.

On 2 November Haitians who believe in voodoo celebrate the Day of the Dead. Followers visit the tombstones of relatives

and pay their respects to Baron Samedi, the god of the dead. To show they are 'possessed', followers often rub hot pepper juice on their bodies. Some hold swearing contests near the gates of the capital's large municipal cemetery.

Confusing world!

Great Danes don't come from Denmark, they come from Germany.

The Red Sea is not red.

The East Alligator River in Australia's Northern Territory harbours crocodiles, not alligators.

The Jerusalem artichoke doesn't come from Jerusalem: it's from America.

French fries originated in Belgium, not France.

Bagpipes were invented in Iran, then brought to Scotland by the Romans.

Venetian blinds were invented in Japan.

Turkish baths originated in ancient Rome, not in Turkey.

Maltesers don't come from Malta.

Petit Suisse cheese is made in France, not in Switzerland.

The French poodle originated in Germany.

Panama hats come from Ecuador – not Panama.

Smallest

At 0.44 square kilometres, Vatican City is the world's smallest country. In fact it's a third of the size of London's Hyde Park. With a population of just over 800, it's the least populated country on earth.

Bermuda is the smallest country in the western hermisphere.

The Gambia is the smallest country in Africa.

Guadeloupe is home to the world's smallest insect.

The world's smallest vertebrate was discovered in Sumatra. A member of the carp family, it grows to a maximum length of one centimetre.

The Falkland Islands have the world's smallest capital city, Port Stanley.

The Maldives is the smallest Muslim nation in the world.

The smallest church in the world is in Kentucky, US. There is room inside for just three people.

Pitcairn Island is the smallest island with country status.

Nauru is the world's smallest island nation.

The world's shortest inter-continental flight is from Gibraltar to Tangiers, Morocco. The flight takes 20 minutes.

The most expensive countries to live in

Japan
South Korea
Russia
Taiwan
Norway
Hong Kong
Switzerland
Denmark
Argentina
China
Finland
Ivory Coast
US
Sweden
Venezuela
UK
Singapore
Oman
Jordan
Kuwait

Words

There's an Albanian insult that translates as 'dwarf with seven tails who can't find rest in his grave'. There is, however, no Albanian word for headache.

The molars that appear between the ages of 17 and 25 are known as 'wisdom teeth' in the UK, but in Romania they're known as 'mind teeth' and in Korea they're called 'love teeth'.

The phrase 'running amok' is derived from the Papua New Guinean word 'amok' meaning 'rage'.

English has borrowed many words from Arabic, for instance: satin, cotton, tariff, checks, saffron, caraway, algebra, alcohol, chemistry and alkali.

Cambodians use a huge number of words in the Khmer language. For example, there are several different ways to say 'eat rice'.

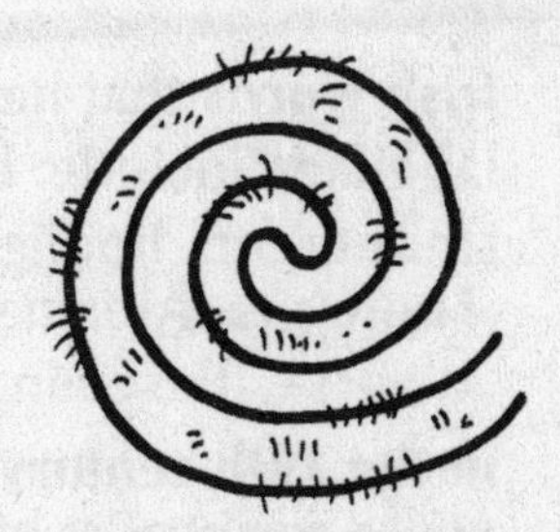

In South Africa the @ is referred to as an 'aapstert', which means a 'monkey's tail'. In Israel it's called a 'strudel', in Slovakia it's known as a 'pickled herring', while in Denmark it's often called 'an elephant's trunk'.

The word 'thug' comes from the Hindu word 'thag' which referred to the religious fanatics who plagued India, carrying out acts of brutality in the name of their goddess.

The Phoenician alphabet, which was in use around 1600 BC, is widely considered the foundation for alphabetic writing in the West and the Middle East.

Irish words that have entered the English language include: banshee, galore, bother, smithereen, hooligan, tantrum and donnybrook.

In the 19th century the French master chef Marie-Antoine Carême created the term 'haute cuisine' (high cooking) for the best French cooking. He also invented the toque, the high white hat that is part of the chef's uniform.

The way other people live (1)

In Negril, Jamaica, buildings mustn't be any higher than the tallest palm tree.

In the fairy tales that Russians tell their children, there's a witch named Baba Yaga, who has iron teeth, flies around in a giant mortar, eats children and lives in a hut that stands on chicken legs.

In Saudi Arabia, the throne passes from brother to brother before going from father to son.

Young girls of the Teda people of the Sahara play with dolls made of mud or wood. The dolls don't have any facial features; instead, little beads are placed in a geometric pattern where the face would be.

In British Columbia, Canada, there's at least one earthquake every day but it's almost always extremely minor.

The San Blas Indian women of Panama consider giant noses a mark of great beauty. They paint black lines down the centre of their noses to make them appear longer.

Each year in Los Angeles people spend 93 hours sitting in traffic on average.

Tanzanians have a special way of measuring time. Sunrise, when the day begins, is 1 a.m. Tanzanian time.

In Vietnam, everyone's birthday is celebrated at the time of the New Year (the same day as the Chinese New Year), or Tet as it's known. The Vietnamese don't know or even acknowledge the exact day they were born. On the first morning of Tet, adults congratulate children on becoming a year older by presenting them with red envelopes that contain Lucky Money. These envelopes are given to the children by parents, siblings, relatives and close friends.

In the Philippines all distances are measured from the Rizal Park in Manila.

In Bolivia, scheduled events usually start late since arriving on time is not expected.

Fish fighting is very popular in Laos. The fish are kept hungry and then put in a tank together where they fight.

Bhutanese consider a clear 'no' to be too blunt: it's not good manners. So they've perfected various ways of saying yes – from full agreement, through 'I'm not sure', to 'no', which is expressed with nothing stronger than 'perhaps'.

In Chad – as in so many countries in the developing world – families consider children to be insurance for their future, and esteem those who have large families. When people die without having had children, they are said to have 'died twice'.

The gauchos of South America use throwing weapons called bolas to hunt birds and animals. Bolas (also known as 'boleadoras') are made of interconnected cords with weights on the ends. The cords twist around the legs of birds or animals so they can't get away.

It's common for Czechs to share a table at a bar or café. If there is only one diner in a restaurant, then someone else will usually join.

For centuries, upper-class Vietnamese girls had their teeth blackened at the age of 12 or 13. White teeth were thought to be vulgar, and blackening was believed to prevent tooth decay. The practice stopped when Vietnam was colonized by the French, who found black teeth unattractive.

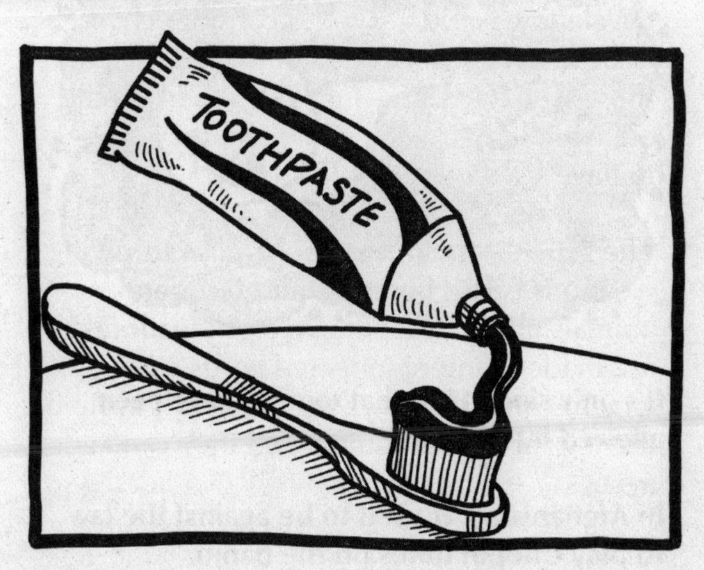

Fascinating facts! (2)

In ancient Japan, public contests were held to see who could break wind loudest and longest. Winners were awarded prizes.

It's only since 1974 that tourists have been allowed into Bhutan.

In Afghanistan, it used to be against the law to play Chopin tunes on the banjo.

There's an elephant orphanage in Sri Lanka.

In the town of Vulcan (in Alberta, Canada), the tourist welcome sign is written in both English and Klingon (the alien language in *Star Trek*).

There's an island in Thailand named after James Bond.

In 1657 a Japanese priest decided to burn a kimono because he thought it carried bad luck. A fire resulted that led to the destruction of over 10,000 buildings and the death of a 100,000 people. Guess the priest was right: the kimono *was* unlucky.

In the 1930s a cult started up on the South Pacific island of Tanna in Vanuatu over cargo brought over from the US. The locals thought that the cargo – labelled John Frum Cargo – had been sent by the gods and started worshipping John Frum. Mind you, there are other people in Vanuatu who worship Prince Philip, the husband of our Queen.

In the 19th century a Frenchman named Casimir Polemus survived three shipwrecks (the *Jeanne Catherine,* which was wrecked off Brest on 11 July 1875; the *Trois Frères,* was wrecked in the Bay of Biscay on 4 September 1880, and *L'Odéon,* wrecked off Newfoundland on 1 January 1882. What makes this even more extraordinary is that, in each case, he was the only survivor.

There is a town in the Cayman Islands called Hell. They have a post office there so tourists can send a postcard from Hell. There's also a town in Norway called Hell.

Countries with more sheep than people

Australia
Falkland Islands
Iceland
Ireland
Mauritania
Mongolia
Namibia
New Zealand
Turkmenistan
Sudan
Uruguay

Countries with more pigs than people

Denmark (13.5 million pigs; 5.5 million people)
Samoa
Wallis and Futuna Islands
Tuvalu

Countries with more cows than people

Argentina
Australia
Botswana
Brazil
Ireland
Mongolia
Namibia
New Zealand
Paraguay
Uruguay

Crime & punishment

In the 17th century, people who were deemed to be enemies of the British crown were sent to Barbados as servants. This practice was so widespread that the punishment was sometimes described as 'being Barbadoed'.

The Chinese have used fingerprints as a method of identification for 1,500 years but it was a Belgian who just used them for forensic purposes.

During the 1920s there was a law in Russia that all private automobiles (i.e. those not used by the Communist government) had to have a yellow stripe painted all the way round them.

It is illegal in Arizona to hunt camels.

Anyone caught drunk in public in ancient China was put to death.

In Belgium, it's against the law for children

to chuck bananas at policemen on Christmas Eve and for policemen to chuck bananas at children on Christmas Day.

Chewing gum is illegal in Singapore.

98 per cent of software in China is pirated.

In Helsinki, Finland, police don't always give parking tickets: sometimes they deflate a car's tyres instead.

In theory, a Chilean can be imprisoned for not voting in an election.

In 1976 microwave ovens were banned in Russia.

In Antwerp, Belgium, it used to be illegal to walk down the main street if you were wearing a red hat.

In Malta it is illegal to drive a car if you're not wearing a top.

At one time it was against the law to slam car doors in Switzerland.

In 1954 the French town of Châteauneuf-du-Pape enacted a law prohibiting flying saucers from landing within its borders. Crazy, you might think – but it worked!

Hang-gliding is illegal in most Ethiopian national parks because it scares antelopes and when they are scared they cause a lot of damage.

In Saudi Arabia, people who steal can be

punished by having their hands cut off.

Internet sites such as MySpace and Facebook are banned in the United Arab Emirates.

Technically it's illegal to die in the British Houses of Parliament.

There are some strange laws which still in theory apply in Germany. Whether or not anyone takes them seriously is another matter. Here are some of them:

Every office must have a view of the sky.
It is illegal to wear a mask.
A pillow counts as a weapon.

In Thailand, it's illegal to leave the house without wearing underwear.

In France, it used to be forbidden to call a pig Napoleon. It was also illegal to sell a doll without a face.

In Singapore, a fine is issued to people who use public toilets without flushing. Spitting and feeding pigeons are also offences there.

It used to be illegal in Morocco to attempt to sell a date tree as dates were an important source of food for the family.

In 1386, a pig was executed by public hanging in France for the murder of a child. The pig was dressed up in clothes for its execution.

Switzerland has the highest number of reported car thefts while Holland has the most reported burglaries.

America has the highest number of prisoners while China has the highest number of executions.

St Kitts & Nevis are two islands but a single country. It is illegal to swear on Nevis – but it

isn't on St Kitts.

In Qatar, pedestrians have to pay a fine if they don't walk properly on the pavement.

Hundreds of years ago in Japan, anyone attempting to leave the country was executed.

Culture

The entire town of Capena in Italy lights up cigarettes each year at the Festival of St Anthony. Even children!

Kadooment – meaning 'fun and good times' – is Barbados's biggest street carnival, held on the first Monday in August.

Pysanky is the Ukrainian art of egg-decorating, which is especially popular at Easter.

The *cueca*, Chile's national dance, mimics a rooster stalking a hen.

Almost everybody in Vietnam knows – and is able to recite parts of – the classical poem, *The Tale of Kieu*, written by Nguyen Du (1765–1820). The poem is important to the Vietnamese because it expresses the belief that the world is governed by 'Vu Tru' – the universal law of addition and subtraction – evens out. So if, for example, a woman is beautiful, her life will be miserable in order to establish balance.

An important celebration for Mexicans is the Festival of Our Lady of Guadalupe (the Patroness of the Americas), which takes place on 12 December. Some people walk, sometimes for weeks, from their homes to the main cathedral in Mexico City.

In Singapore, many people enjoy bird-singing competitions. The owners of the birds spend a lot of time training them. A bird that sings well can bring in thousands of dollars in prize money. A National Songbird Competition has been held annually since 1982.

'He who brings the kola nut brings life' is a popular expression in many parts of Nigeria. The kola nut is an important symbol in the Nigerian belief system. Breaking the kola nut is part of any welcoming ceremony.

The Malaysian dagger known as the *kris* (or *keris*) is not just a weapon, it's also a spiritual object. The kris, which is shaped like a snake or dragon, is said to give its owner magical

powers.

The limbo dance originated on the island of Trinidad.

Legends

To the ancient Greeks and Romans, Gibraltar was one of the two Pillars of Hercules, set up by the mythical hero to mark the edge of the known world.

The Aztecs believed that they lived in a world in which earlier peoples had been destroyed by the death of the sun. As a result, human sacrifices were made in an attempt to keep the sun alive.

Many Vietnamese people believe that they are the descendants of the Underwater King.

According to Maori legend, New Zealand was first discovered when fishermen were chasing a massive octopus.

According to Australia's Aborigines, the world was originally flat, featureless and grey. Then huge creatures woke up and wandered the Earth. As they hunted for food and dug for water, they created mountains, valleys and rivers. This period is known as

the Dreamtime. Aboriginal life began in the Dreamtime. Once creation was complete, the creatures disappeared.

Legend has it that Sri Lanka was formed when Hanuman, the Monkey King, chucked a mountain into the ocean.

Deepest

The deepest canyon in the world is Hell Canyon in Idaho, US. Its maximum depth is 2,400 metres.

The deepest mine in the world is Western Deep Levels near Carletonville, South Africa. It's 4.2 kilometres (2.6 miles) deep.

The deepest lake on the planet is Lake Baikal in Russia, which goes as deep as 1,680 metres. It is also the world's oldest lake, having been formed about 25 million years ago.

The deepest spot in the ocean (and therefore on Earth) is the Mariana Trench in the Pacific Ocean. Its maximum depth is 11 kilometres or 6.8 miles. Contrast that with the highest point on Earth (the top of Mount Everest), which is 'only' 8.85 kilometres high.

The world's deepest post box is located 10 metres under the sea at Susami Bay, Japan. It is used by divers and emptied daily by the post office.

National anthems

The Vanuatu national anthem is called 'Yumi, Yumi, Yumi'.

5 per cent of Canadians don't know the first seven words of the Canadian national anthem, but do know the first nine of the American anthem.

The Estonian national anthem has the same tune as Finland's but different words.

The national anthem of Greece is 158 verses long. It is said that no one knows all the words by heart.

The national anthem of the Netherlands is the oldest national anthem in the world. The anthem takes the form of an acrostic: the first letters of each of the 15 verses represent the name 'Willem van Nassov', or William of Orange as we'd call him.

Superstitions

In Chile, there are various beliefs associated with New Year's Eve – such as eating lentils at midnight for good luck or walking around with a suitcase in order to increase your chances of travel in the coming year.

Some Burmese people believe that you shouldn't have a haircut on a Monday, a Friday or on your birthday. Bad luck for hairdressers though!

In Russia, it's customary not to shake hands across a doorway as it might lead to an argument.

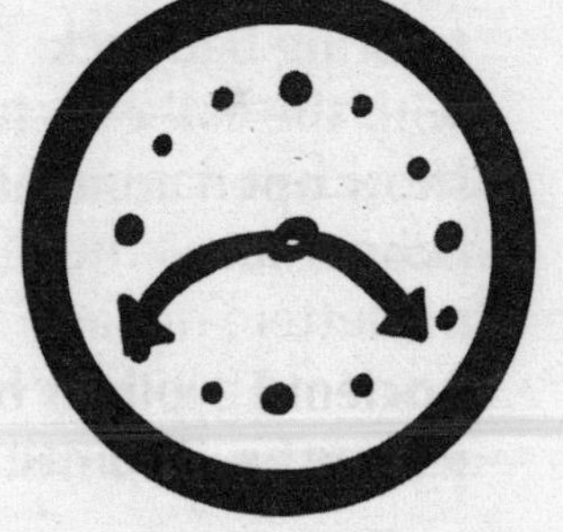

In China, it is considered unlucky to give someone a clock for a birthday present. In Mandarin, the word for 'clock' is similar to the word for death.

There's a fern that grows in the forests of Belarus but rarely flowers. It's said that if you find one that's blossoming, you'll find happiness. It is also said that evil spirits follow people in the forests to stop them finding the magic ferns.

Storks are a common sight in Bulgaria. Although they build untidy nests on roofs and on top of telegraph poles, they're considered to be lucky.

Egyptians fear the evil eye, which is believed to bring bad luck. Envy is often associated with the evil eye. It is considered rude to show open admiration for another person's possessions.

Ancient Egyptians believed that onions could fight off evil spirits.

Superstitious Australians say you must always make eye contact when toasting someone, otherwise you will suffer seven years' bad luck (like breaking a mirror).

Koreans never use red ink when writing a person's name as it signifies that they believe that the person is dead – or, worse, that they want them dead.

The Turks have many superstitions. They believe that if they see a spider somewhere in their homes – or drop food on their clothes while eating – they'll have some guests soon. When they see a black cat, they take hold of a lock of their hair for fear of bad luck. They also have a wonderful superstition that if you stand between two people whose names are the same, you should wish for something because your wish will come true.

Kazakhstanis never whistle inside their own house – or anyone else's – for fear that it will bring poverty to the owner of the house.

The Chinese don't give knives, scissors or any other cutting instruments to one another as they indicate a desire to end the relationship. Giving clocks, handkerchiefs or straw sandals are also taboo as they're associated with death or funerals.

In Venezuela, it is said that if someone passes a broom over your feet, you will never get married.

The Japanese believe that if you put a piece of snake skin into your wallet, you will become rich or find money. This is because snakes are a symbol of money and wealth in Japan. Some people believe the snake is an animal of God, so they never kill snakes – for fear of losing their money.

In Ukraine, it's considered good luck to find a spider's web in the house on Christmas morning.

Place names

There's a province in Turkey named Batman. It got its name as a shortening of the Bati Raman mountains.

Himalaya means 'home of snow'.

Erik the Red was the Viking explorer who first discovered Greenland. There are a number of theories as to why he named it Greenland when it was (and still is) mostly ice. Some say that it was to lure settlers away from Iceland (which was, just to confuse you, mostly green) and persuade them to settle in this icy region. Other people claim that, at the time of its discovery, Greenland was a lot more green than it is now.

There is an alternative explanation: maybe Erik the Red just wasn't very perceptive.

The coastal region of Croatia is called the Dalmatian Coast and this is where the Dalmatian dog comes from.

To the Mexicans, Mexico really means the capital, Mexico City. Mexico itself is referred to as La República Mexicana.

The capital of Christmas Island is named Flying Fish Cove.

Christopher Columbus named Costa Rica (rich coast) under the misapprehension that the land was filled with precious metals.

The name Cuba comes from the Taíno word *cubanacán*, meaning a centre or central place.

There's an island in the Seychelles called Silhouette.

Barbados was given its name by a group of Portuguese sailors who landed on the island on their way to Brazil in 1536. Here they found huge banyan trees with roots hanging like beards from the branches. They called the island 'Los Barbados' meaning 'the bearded ones'.

The capital of Anguilla is, quite simply, The

Valley. In Italian and Spanish, *anguilla* means *eel*. This is fitting because the island has a long, thin shape.

Argentina means 'Land of Silver', which is odd because there's hardly any silver there, contrary to what the first explorers must have thought.

Ankara in Turkey used to be known as Angora. The goats in this area produced the famous fine angora wool known as mohair.

The literal translation of Belarus is 'White Russia'.

The Republic of Bolivia was named after Simón Bolívar- the man who liberated Bolivia from Spanish rule.

Tierra del Fuego (Land of Fire), a region of South America divided between Argentina and Chile, was given its name by Ferdinand Magellan, the Portuguese explorer who, in

1520, became the first European to discover it. He was astonished by all the fires he saw and thought that the locals were waiting in the forests to ambush his ships. It turned out that the fires had been lit by the Yamana Indians simply to ward off the cold!

Christmas Island was given its awkward name because the explorers who landed there in 1643 arrived on Christmas Day. As it's in the southern hemisphere, Christmas on Christmas Island is in the middle of summer.

The name of Canberra, Australia's capital, comes from the Aboriginal word for meeting place. It was chosen as the capital because it's just about equidistant between Sydney and Melbourne.

The ancient Greeks named Malta 'Melita', meaning 'Island of Honey'.

The name of the Honduran capital city, Tegucigalpa, comes from two indigenous words: *teguz*, which means 'hill', and *galpa*, which means *silver*. Tegucigalpa was once a silver mining town.

Long place names

There is a place in Wales that has the longest officially recognized place name in the United Kingdom: Llanfairpwllgwyngyllgogerychwyrn-drobwllllantysiliogogogoch. This translates into English as 'the church of St Mary in the hollow of white hazel trees near the rapid whirlpool by St Tysilio's of the red cave'.

Llanfairpwllgwyngyllgogerychwyrn-drobwllllantysiliogogogoch.com is the longest single word (without hyphens) .com domain name in the world.

Moretonhampstead in Devon boasts the longest one-word place name in England.

Taumatawhakatangihangakoauauotamatea-pokaiwhenuakitanatahu is the name of a hill in the Hawke's Bay region of the North

Island of New Zealand. At 85 letters, this is the longest geographical name in the world. It translates into English (from Maori) as 'the summit where Tamatea, the man with the big knees, the climber of mountains, the land-swallower who travelled about, played his nose flute to his loved one'. The locals, not surprisingly, call it simply Taumata.

The full ceremonial name of Bangkok is in fact the world's longest place name. In full, it's Krung Thep Mahanakhon Amon Rattanakosin Mahinthara Yuthaya Mahadilok Phop Noppharat Ratchathani Burirom Udomratchaniwet Mahasathan Amon Piman Awatan Sathit Sakkathattiya Witsanukam Prasit. Thai children have to learn this and recite it in school. Poor them!

In Webster, Massachusetts, there is a lake called Chargoggagoggmanchauggagogg-chaubunagungamaugg. It's also known as Webster Lake.

There's a village in Ireland called Muckanaghederdauhaulia. In English this means 'pig-marsh between two seas'.

The richest countries in the world

(calculated by how much each inhabitant would get if all the money in the country were shared out equally)

Luxembourg: £34,400
United Arab Emirates: £24,850
Norway: £23,800
Ireland: £21,800
United States: £21,750
Andorra: £19,400
Denmark: £18,500
Austria: £17,800

The poorest countries in the world

(calculated by how much each inhabitant would get if all the money in the country were shared out equally)

Comoros: £300
Malawi: £300
Solomon Islands: £300
Somalia: £300
Burundi: £350
Democratic Republic of the Congo: £350
Afghanistan: £400
East Timor: £400
Tanzania: £400
Yemen: £450

Flags

The Isle of Man's flag is three legs joined together.

The colours of Jamaica's flag are symbolic. Black signifies the strength and creativity of the people; gold, the natural wealth and beauty of sunlight; and green, hope and agricultural resources.

The cedar tree on Lebanon's flag symbolizes strength, holiness and eternity. The white background stands for peace, and the red bands represent sacrifice.

Libya has the only flag that's just one colour with nothing else on it. The one colour is green and it represents Islam.

Mozambique and Guatemala are the only two countries with flags that have guns or rifles on them.

The Nepalese flag is the only flag that's not four-sided. It takes the form of two overlapping triangles on top of each other.

The Philippines is the only country in the world that holds its flag upside down when at war: the blue part is displayed at the top in times of peace whilst the red part is at the top in wartime.

The Ukrainian flag of blue and yellow symbolizes a field of yellow grain with a blue sky overhead.

Work

Around the world – especially in developed countries – robots do a lot of the grunt work in factories. They cost less than people and can work in unhealthy environments. There are also factories in Japan where robots are working on creating and constructing more robots.

One-third of El Salvadorans live in San Salvador. Because of poor economic conditions in the countryside, many parents have to leave their children in the care of grandparents in order to search for work in the city.

10 per cent of all Americans earn their

living in the motor industry or in a company closely associated with the motor industry.

About 10,000 Albanians make their livelihood hunting frogs. Albania exports 400 tons of live and frozen frogs every year, mostly to France and Italy, where frog's legs are considered a delicacy.

Zambia is one of the world's largest exporters of roses and many Zambians work in the rose business. Zambia can grow them throughout the year and sell them for a low price. The demand for roses within Zambia is low, as it isn't customary to give flowers there.

The cities with the largest populations

Tokyo (Japan): 33,750,000
Mexico City (Mexico): 21,850,000
New York (US): 21,750,000
Seoul (South Korea): 21,700,000
São Paulo (Brazil): 20,200,000
Mumbai (India): 18,800,000
New Delhi (India): 18,100,000
Los Angeles (US): 17,450,000
Osaka (Japan): 16,700,000
Jakarta (Indonesia): 16,300,000
Cairo (Egypt): 15,600,000
Moscow (Russia): 15,350,000
Calcutta (India): 14,950,000
Manila (Philippines): 14,000,000
London (UK): 13,900,000

That's surprising! (3)

Construction in Dubai has reached such a high level that a quarter of the world's construction cranes are there.

The Hungarian capital city of Budapest was once two separate cities: Buda and Pest.

In Australia, a credit card was issued to a cat named Messiah. Its owner wanted to test the bank's security system.

Guam's beaches have ground coral instead of sand.

When English-language films are dubbed in Poland, all the voices are done by a male actor. So the viewer ends up watching women and children talking with a man's voice.

Old-fashioned Chinese typewriters have 5,700 characters.

Although Greenland is part of the continent of North America, it's a Danish province (though these days it's self-governing like Australia and New Zealand are).

Croatia has more than 1,000 islands.

There's a lake high up in the mountains of Kyrgyzstan called Lake Issyk-Kul, where the water never freezes despite the altitude.

Ethiopia celebrated the dawn of the new millennium in September 2007.

Many scenes in *Star Wars* were filmed in Tunisia. When filming first started, the country experienced its first rainstorm in over 50 years. During the filming, a massive military vehicle from the set was stationed on the border with Libya – causing the Libyan government to think that the Tunisians were launching an invasion.

The largest lake in the world is the Caspian Sea.

Only a tenth of Greece's islands are inhabited.

On 9 October, *Hangul* is celebrated in South Korea. This is a day for celebrating the alphabet.

The Solomon Islands are home to an active underwater volcano.

In Taiwan, it is the Year 100 in 2011.

The Tonle Sap River in Cambodia flows north for one half of the year and then south for the other half.

The busiest underground (subway) systems in the world

Tokyo: 2.9 billion passenger rides a year
Moscow: 2.5 billion
Seoul: 1.6 billion
New York City: 1.5 billion
Mexico City: 1.4 billion
Paris: 1.4 billion
Hong Kong: 1.2 billion
London: 1 billion
Osaka: 880 million
São Paulo: 845 million
St Petersburg: 830 million
Shanghai: 814 million
Santiago: 803 million
Beijing: 765 million
Cairo: 750 million

Nicknames

Barbados's nickname is 'Little England'.

La Paz, the capital of Bolivia, is nicknamed 'The City That Touches the Sky'.

The Malaysian city of Ipoh – 125 miles north of the capital, Kuala Lumpur – is nicknamed 'The City of Millionaires'.

Montserrat is nicknamed 'The Emerald Isle of the Caribbean' because of its similarity to Ireland.

Victoria Falls – on the Zambezi River between Zambia and Zimbabwe – is known locally as Mosi-oa-Tunya: 'The Smoke That Thunders''.

People who live in Kuala Lumpur are known as KLites.

The region in which Morocco, Algeria and Tunisia are located is known in Arabic as the Maghreb or 'sunset'. It got this name

because it lies so far to the west of other Arab nations.

One-seventh of Sweden lies north of the Arctic Circle. This region is called the Land of the Midnight Sun because the sun shines almost 24 hours a day for several weeks in the summer months of June and July. Unfortunately for them, during the winter months of December and January, the sun barely rises above the horizon.

The US has the nickname 'Uncle Sam'. This comes from a New York meat inspector called Uncle Sam Wilson.

The Moroccan city of Marrakesh is called the 'Red City' because of the colour of its buildings, which were constructed using the red sand found in that region.

Bohemia is sometimes called 'The Roof of Europe'. Interestingly, no rivers or streams flow into the country, but many rivers flow out of it.

Captain Cook nicknamed the Tongan islands 'The Friendly Islands' because of the warm reception the locals gave him upon his arrival.

Countries and their most popular surnames

Argentina: Fernández
Belgium: Peeters
China: Li
Denmark: Jensen
Estonia: Oak (many popular Estonian surnames have something to do with nature)
France: Martin
Greece: Papadopolous
Hungary: Nagy
Ireland: Murphy
Japan: Sato
Lithuania: Kazlauskas (men); Kazlauskien (women)
Mexico: Hernández
The Netherlands: De Jong
Poland: Nowak
Russia: Smirnov
Spain: García
Taiwan: Chen
UK: Smith
Vietnam: Nguyen

Money

Australia pioneered the use of bank notes made of plastic (polymers). They last four times as long as regular paper notes and provide greater security against counterfeiting.

In 2008, inflation in Zimbabwe went over two million per cent.

The US has more billionaires than any other country in the world.

In Croatia in 1994, the kuna replaced the dinar as the national currency. The Croatian kuna is an indigenous animal whose pelt was used as a means of currency during the Middle Ages.

The Brazilian real is an interesting-looking currency because on one side the image is horizontal while on the other side it's vertical.

Until the 20th century, some Ethiopians still

used an ancient form of currency called amole, which are salt bars. The system arose because salt was scarce.

Malaysia's currency is called the ringgit which is the Malay word for 'jagged'. This is supposed to be a reference to the jagged edges of the Spanish silver dollars that were used in this region.

Tulip bulbs were a form of currency in the 17th century in the Netherlands.

Until the 19th century, blocks of tea were used as money in Siberia.

Cocoa beans were used as money in ancient Guatemala. Counterfeiters were at work even in those days: some people removed the insides of the beans and filled them with clay.

In Togo, it's customary to use the right hand when giving or receiving money.

The Rough Guide's 25 wonders of the world

Salt flats of Salar de Uyuni, Bolivia
Uluru or Ayers Rock, Australia
Pyramids at Giza, Egypt (the only one remaining of the Seven Wonders of the World)
Drifting down the Amazon
'Fairy chimneys' and caves of Cappadocia, Turkey
Grand Canyon, Arizona, US
Petra, the city carved from stone in the Jordanian desert
Machu Picchu, Peru
Sagrada Familia, Gaudí's masterpiece in Barcelona
Perito Moreno glacier, Patagonia
Sistine Chapel, Rome
Trekking in the Himalayas
Angkor Wat, Cambodia
The canals and palaces of Venice
Taking a camel train across the Sahara
Great Wall of China
Victoria Falls, Zambia and Zimbabwe
Paddling in the barrier reef, Belize

Taj Mahal, India
Maya ruins of Mexico and Guatemala
Stone giants of Easter Island
Grand Mosque, Djenné, Mali
Gambling in Las Vegas
Forbidden City, China
Itaipú, the world's biggest dam, Paraguay and Brazil

I would add the following sites and experiences to that list:

Tower of London
Colosseum, Rome
Niagara Falls, Canada and the USA
Northern Lights
Great Barrier Reef, Australia
Sequoia giant redwood trees of California
Granada's Alhambra in Spain
Eiffel Tower, Paris
Burj Al Arab in Dubai
Jerusalem's old city
Ruins of Pompeii and Herculaneum, Italy
Winter Palace, St Petersburg, Russia
Tivoli Gardens, Copenhagen
Santorini – and the clifftop village of Oia.

Geysers and thermal pools in Rotorua in New Zealand
Parthenon, Athens
Disney World, Florida
A herd of migrating wildebeest in the Masai Mara (preferably viewed from a hot-air balloon)
Times Square, New York City
Universal Studios, California

The world we live in

The geographical centre of Europe is Slovakia – to be precise, the Krahule hill near the mining town of Kremnica.

Slovenia is situated at the crossroads of central Europe, the Mediterranean, and the Balkans.

Some 70 million people across the globe have Irish ancestors.

The Atlantic Ocean is saltier than the Pacific.

There are over 4,000 caves in Bulgaria.

Peat bogs cover more than 15 per cent of Ireland.

Stockholm is built on 14 islands which are all joined together by bridges.

More than half of Saudi Arabia is desert.

Madrid, the capital, is in the precise centre of Spain.

Vietnam is shaped like a massive 'S'.

Mali is shaped like a butterfly.

And then there's the Bermuda Triangle . . . an area of the North Atlantic, between Bermuda, Puerto Rico and Florida. This has been the site of numerous shipwrecks, plane crashes and other mysterious disappearances. And there are some strange theories. These include UFO landings; space–time warps; unusual levels of magnetic power; and the side-effects of an experiment gone wrong. The intriguing and dangerous effects of the Bermuda Triangle were noticed a long time ago, when early explorers nicknamed Bermuda 'The Isle of Devils' due to the effect it had on ships. In fact, there is no Bermuda Triangle: every accident can be rationally explained – although that hasn't stopped superstitious people believing in it.

Oldest

Dolní Věstonice, a small village in the South Moravian region of the Czech Republic, is the most ancient village in the world. It's reckoned to be 27,000 years old.

San Marino is the world's oldest republic, having been founded in the year 301.

Bhutan is the world's newest republic, having been founded in the year 2008.

The world's oldest vine – some 400 years old – can be seen growing in Maribor's old town in Slovenia.

The Namib Desert in Namibia is the oldest desert in the world.

Ethiopia is the oldest country in the world – dating back to 3000 BC. Montenegro was made independent from Serbia in 2006, making it the world's newest country.

As of 2008, Nepal is the most recent country to be made a republic.

South Africa has the world's oldest meteor crater. It's around two billion years old.

The ancient city of Byblos in Lebanon, known today as Jbeil, is the world's oldest continuously inhabited city. It's mentioned in the Bible in 1 Kings 5:18, referring to the nationality of the builders of Solomon's temple. Its name comes from the Greek word for papyrus (*bublos*) as it was a stopping point for papyrus shipments from Egypt to Greece. The Greek word *biblion* (book) is the origin of the English word Bible. During the Crusades, Byblos was referred to as Gibelet.

The oldest man ever (provable!) was a Japanese man named Shigechiyo Izumi, who died in 1986 at the age of 120 years and 237 days. He also holds the record for the longest working career, having worked for 98 years. He was just 1.42 metres tall, weighed 42.6 kilograms and lived through 71 Japanese Prime Ministers.

Just as impressive, at the age of 98, Dimitrion Yordanidis of Greece could run a 26-mile marathon (in seven hours, 33 minutes) – making him the oldest marathon runner in history.

Syria boasts the oldest civilization still in existence today and Damascus is the oldest capital in the world (founded in 2500 BC).

Syria also boasts the world's oldest alphabet.

The oldest zoological garden is in Vienna, Austria (founded in 1752).

The oldest human skull in the world was found in Chad in 1960. The remains were found close by the teeth of an extinct elephant and are thought to be 200,000 to 500,000 years old.

I said we ar-
closed today

Dubai is home to the world's most advanced artificial ski centre. It's just under 3,000 square metres and they use real snow. It even snows inside.

Fiji is right on top of the International Date Line, which has been bent round to accommodate Fiji's understandable desire to have its whole country living in the same day. When I went to Fiji, I flew home via Los Angeles. Because Fiji is a whole day ahead, we arrived in LA 12 hours before we'd set off – which was more than a little confusing!

A Russian scientist tried to cross-breed humans with apes to create the Humanzee. This was, fortunately for both species, unsuccessful. However, another Russian scientist did succeed in creating a two-headed dog.

Graça Machel, born in Mozambique, married the presidents of two different countries. She was married to Samora Machel, the President of Mozambique, until his death in 1986. Then, in 1998, she married Nelson Mandela, the President of South Africa.

People who make world maps hate Montenegro because it's so hard to fit the whole name of the country into such a small space on the map.

Since the 18th century it has been illegal for the bull to be killed in Portuguese bullfighting – while in Costa Rican bullfights, the bull often wins.

Montezuma, an Aztec emperor of Mexico, drank up to 50 glasses of chocolate a day.

There was a Zambian president who threatened to resign unless the people stopped drinking so much alcohol.

In 1969 there was a week-long war between El Salvador and Honduras. It was called the Soccer War because it started at a football match. Of course, the football riot was just the trigger for tension that had built up between the two countries for a couple of years. But a lot of people thought that the war was just over a football match. Incidentally, it was a World Cup qualifying match which El Salvador won. It wasn't worth a war, though,

because they both got knocked out in the first round of the 1970 World Cup.

Scientists at the University of Guam have discovered that fish can fart. The fish, on the other hand, knew it all along – which suggests that they're cleverer than the scientists.

Transport

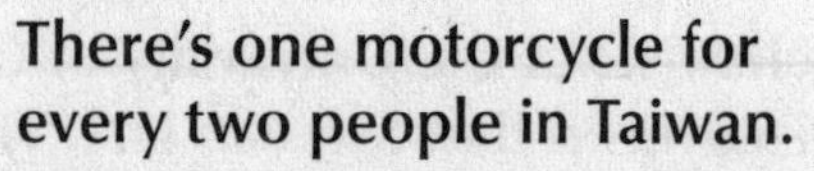

There's one motorcycle for every two people in Taiwan.

More than 20 million people ride bicycles in Vietnam.

Until 1965, Sweden drove on the left. The conversion to driving on the right was made on a weekday at 5 p.m. The late afternoon was judged to be the best time of day to avoid accidents.

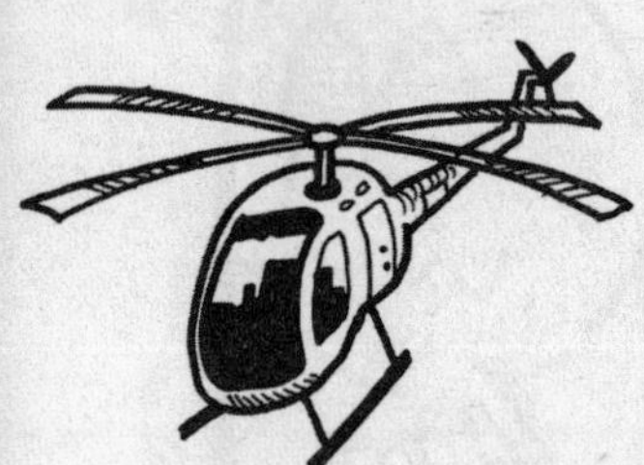

The world's longest limousine is over 14 metres long and has a built-in swimming pool. It costs more than £3,000 a day to rent it.

There's one car or lorry for every 10 people in the world.

There are 400 million car owners in the world. There are twice as many bicycle owners.

Jeepneys are the most popular means of public transport in the Philippines. The jeepney is a small truck with fancy bodywork covered with paintings, flags and flowers. Originally created from leftover American jeeps, jeepneys are now made in factories and then colourfully customized by their owners.

There are no airports in Monaco (the nearest is Nice in France).

With 16 million bicycles, the Netherlands has more per person than any other country in the world. In fact, the Dutch have enough bicycles for every single member of the population to own one.

In 1928 Richard Halliburton swam the Panama Canal. He had to pay a toll of 36 cents.

In California, a driverless car has been developed. It's driven by a robot with laser eyes and an electronic brain. You'd be very frightened indeed if you saw one of those coming towards you . . .

Countries with unoriginally named capital cities!

Algeria – capital: Algiers
Andorra – capital: Andorra La Vella
Brazil – capital: Brasilia
Djibouti – capital: Djibouti
El Salvador – capital: San Salvador
Gibraltar – capital: Gibraltar
Guatemala – capital: Guatemala City
Guinea-Bissau – capital: Bissau
Kuwait – capital: Kuwait City
Luxembourg – capital: Luxembourg
Mexico – capital: Mexico City
Panama – capital: Panama City
San Marino – capital: City of San Marino
Singapore – capital: Singapore City
Tunisia – capital: Tunis
Vatican City – capital: Vatican City

Schools & education

There are eight countries in the world which can claim 100 per cent literacy among their population. They are Andorra, Finland, Norway, Greenland, Georgia, Liechtenstein, Luxembourg and the Vatican City.

Together with Britain and the US, there are 44 countries in the world where the literacy rate is above 99 per cent.

At the other end of the scale, there are 20 countries in the world where under half the population is literate. Almost all of them are in Africa, though they also include Pakistan and Afghanistan – the latter boasts the second-lowest literacy rate of just 28 per cent. Burkina-Faso is the lowest of all, with just 21 per cent of the population who are literate.

For children in Pakistan, primary education is not compulsory. In the cities, some children are expected to find jobs and work to help the family.

In Bahrain, boys and girls go to separate schools. When there is only one school in a village, boys go to school in the morning and girls go in the afternoon.

The world's average school year is 200 days. In the US it is 180 days; in Sweden 170 days. In Japan, however, the school year is 243 days long – making it comparable to the length of the average worker's year.

Children in Trinidad and Tobago are legally required to go to school for just six years – between the ages of six and twelve.

Students from all over the world go to Montreal in Canada to learn to be acrobats, trapeze artists, dancers and actors at the École Nationale du Cirque. This circus school works closely with the world-famous Cirque du Soleil.

Some of Brazil's main cities have night schools for the poor which teach, feed and provide medical care for children.

Education in Vietnam is often by rote (i.e. children repeat facts given to them by their teachers aloud until they've learned them off by heart – the way some of us learned our tables). In addition, Vietnamese students are expected to stand when they answer questions.

To help Zambia feed itself, the government requires all schools to have gardens. Students learn how to grow fruits and vegetables.

In Vietnamese schools, gongs are used instead of bells to signify the end of lessons, etc.

A good African hint for saving pencils: cut your pencil into three small pieces. Take one piece and wear it on a string around your neck. That way if you lose your pencil it'll only be a third of it.

In Lithuania, on the first day of the school

year in September, young children traditionally bring bouquets of flowers to their teachers.

Most secondary education in Bhutan is taught by Buddhist monks.

In Malaysia, the state schools have to accommodate so many students that some schools operate a shift system. The first shift runs from 9 a.m. to 1 p.m. and the second from 2 p.m. to 6 p.m.

The world's largest school is in the Philippines. Around 20,000 pupils are taught there.

Birthdays

A birthday tradition in Nepal is to place a mark on the birthday child's forehead for good luck.

In parts of Canada – especially along the Atlantic Coast – they have a tradition whereby the birthday child is 'ambushed' and their nose is greased for good luck. Apparently, the greased nose makes the child too slippery for bad luck to catch them. This tradition is thought to be of Scottish origin.

Instead of a birthday cake, many Russian children get a birthday pie with a birthday greeting carved into the crust.

In Norway, the birthday child stands in front of their class and chooses a friend to share a little dance while the rest of the class sings 'Happy Birthday'.

Children celebrating their birthday in Brazil get sweets shaped like fruit and vegetables. Because these sweets are so beautifully made, the children try wait for a while before they eat them. Meanwhile, their houses are decorated with festive banners and brightly coloured paper flowers.

Argentinians pull once on the earlobes of the birthday boy or girl for each year of their age.

In Aruba, the birthday child takes a treat to school for their classmates and teachers. Each teacher also gives the birthday child a small gift like a pencil, an eraser or a postcard. The birthday child is also allowed to wear their 'own' clothes instead of the school uniform.

Mexican birthday celebrations include the piñata, a brightly coloured paper container filled with sweets and sometimes toys. The piñata is suspended from a tree branch or from the ceiling and the children use sticks to try to break it open so that they can get at the sweets and other goodies.

Onlys (2)

Brazil is the only country which is crossed by both the Equator and the Tropic of Capricorn.

Australia is the only country that's also a continent (and it's also the world's smallest and driest inhabited continent). Australia is also the only continent without active volcanoes or glaciers.

The aye-aye, a primate with distinctively long fingers, the tail of a squirrel, the ears of a bat and the eyes of a cat, can only be found in Madagascar.

The only country with a national dog is Holland (the keeshond).

Australia boasts the only bird – the Australian mound-builder bird – that can fly the moment it hatches.

Australia has the world's only wild camels.

Only two countries have borders on three oceans – the US and Canada.

Belize is the only country in the world with a jaguar reserve. Belize is also the only Central American country where English is the official language.

The only British soil occupied by German troops in the Second World War was The Channel Islands.

Gibraltar is the only place in Europe where you can find wild monkeys.

Greece and Australia are the only two countries to have participated in every single one of the Modern Olympic Games.

Malayalam, spoken in Kerala, southern India, is the only language with a palindromic name

– which means it reads the same backwards.

Lebanon is the only country in the Middle East that doesn't have a desert.

Liechtenstein and Uzbekistan are the only two doubly landlocked countries in the world, i.e. they're both entirely surrounded by landlocked countries.

Lemurs can only be found in Madagascar.

The Italian city of Venice is the only city that lies on hundreds of small islands. People travel between buildings in boats.

The only freshwater sharks in the world live in Lake Nicaragua. They are thought to be the descendants of saltwater sharks that were trapped there when the lake separated from the ocean, and they then evolved to survive in a freshwater environment.

Istanbul is the only city in the world that spans two continents. Most of Istanbul lies in Europe but its suburbs are in Asia.

People & places

People from Trinidad and Tobago are sometimes known as Trinbagonians.

If you read the word 'ghost', what image does it conjure up for you? Chances are you'll say a white, misty figure because that's how most of us imagine ghosts. However, in the Caribbean, ghosts – or 'jumbies' as they're known – are imagined as dark shadowy figures.

In Lithuania, forests play an important part in folk tales. During times of war, the forests were a safe haven for those in danger. The oak tree was worshipped during pre-Christian times and today represents longevity and strength. Lithuanians often plant oak trees to mark important occasions.

Robert Louis Stevenson, author of *Treasure Island* among other classics, was buried in Samoa. He had lived there for the last few years of his life. The Samoans referred to him as their 'storyteller'.

Talking of great people, let me introduce you to Maurice Benyovszky, a Hungarian count, who lived for just 40 years in the 18th century (1746–1786). But what a lot he crammed into that short lifetime! At various times he was an explorer, a writer and a chess player. He spoke five languages fluently. He was a French colonel, a Polish military commander and an Austrian soldier. Oh yes, and he also became the king of Madagascar.

Many people in India believe the Ganges River has the power to heal the sick. As a result, millions of people live along its banks and bathe in it.

No one in the Faroe Islands lives more than three miles away from the sea. That's because nowhere on the Faroe Islands is more than three miles away from the sea!

The Andes have more people living in them than any other mountain range in the world.

In Papua New Guinea, less than one-fifth of the population lives in a town or city.

Privacy is very important to the Saudi Arabians, which is why they build their houses with big walls. Guests visiting a Saudi Arabian house are advised to stand at the door in such a way that they can't see inside and only enter when their host gives the signal – by extending his right hand, palm up, and telling them to come in.

Those born in San Marino remain citizens and can vote no matter where they live.

Greenland is one of the most sparsely inhabited areas in the world, with only about 150 towns and villages. There is no railway, few roads, and not many cars.

For a long time the people of Sweden didn't have a national day. So, as recently as 1983, 6 June was named the official national day. Although it remains a regular workday, some towns have parades, and people display the Swedish flag.

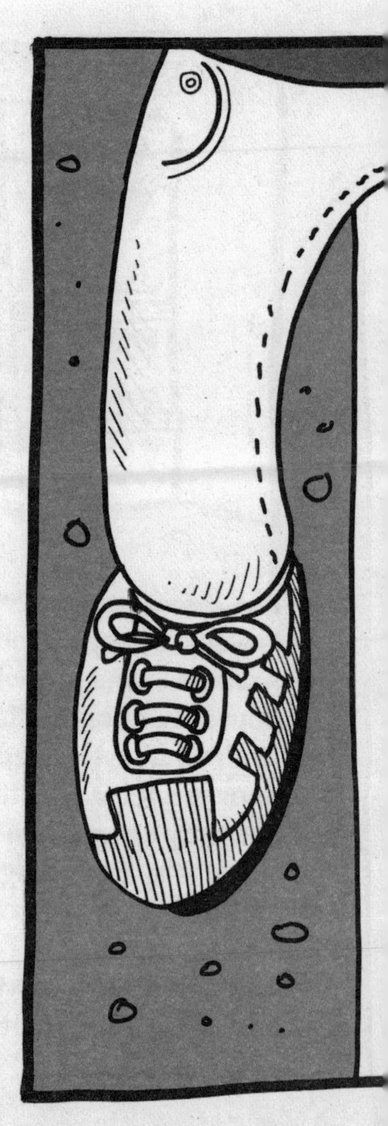

Ecuador is Spanish for 'Equator' – which passes straight through this country. In fact, many visitors come to Ecuador to stand with one foot in the northern hemisphere

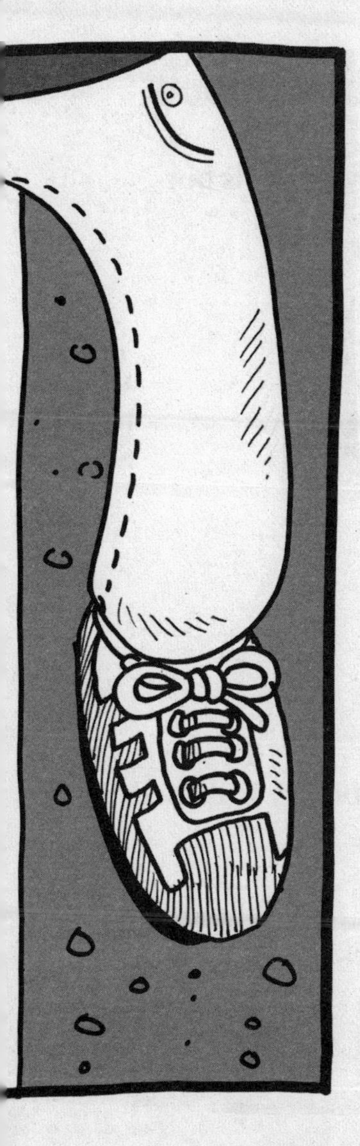

and the other foot in the southern hemisphere.

Thingyan, the Burmese new year, falls some time in mid-April every year. Like Easter, with which it often coincides, it has no fixed date (the date is calculated according to the traditional Burma lunisolar calendar). It's Burma's most important public holiday and, to celebrate, people throw water at each other, using buckets, water-pistols or any other water-delivering devices they can lay their hands on.

Countries and their symbols

Some countries have many symbols but here's a selection:

Australia – Kangaroo
Belgium – Red poppy
China – Dragon
Denmark – Beech tree
Ecuador – Galapagos tortoise.
France – Rooster
Greece – Olive branch
Ireland – Shamrock
Mexico – Dahlia
Netherlands – Tulip
Portugal – Cockerel
Russia – Bear
Scotland – Thistle
Thailand – Elephant (the country is shaped like an elephant's head).
US – Bald eagle
Vietnam – Water buffalo
Wales – Daffodil, leek, dragon

Weather

Mali is the hottest country on Earth.

Greenland is the coldest country.

The Atacama Desert in Chile is the world's driest place. No rainfall has ever been recorded there.

Nights in the tropics are warm because moist air retains the heat well. Desert nights, on the other hand, get cold rapidly because dry air doesn't hold heat to the same degree.

Japan is hit by up to 30 typhoons a year. Winds can reach 200 kilometres an hour and 30 centimetres of rain can fall in 24 hours.

Russia 'boasts' the lowest recorded temperature in Europe: minus 68 degrees Celsius in 1933.
Morocco has the lowest recorded temperature in Africa: minus 24 degrees Celsius in 1935.

The River Nile, the world's longest river, has frozen over twice – once in the ninth century, and again in the 11th century.

Iraq and the Persian Gulf states have two important winds. The eastern *sharki* wind is hot and humid, while the northern *shamal* wind brings welcome cooler air during the hot summer.

In Bosnia and Herzegovina, the *jugo* is a wind that brings rain to various parts of the country.

This is one of those strange-but-true facts worth noting: Argentina not only has the lowest recorded temperature in South America (minus 33 degrees Celsius in 1907), it also has the highest recorded temperature in South America (49 degrees Celsius in 1905).

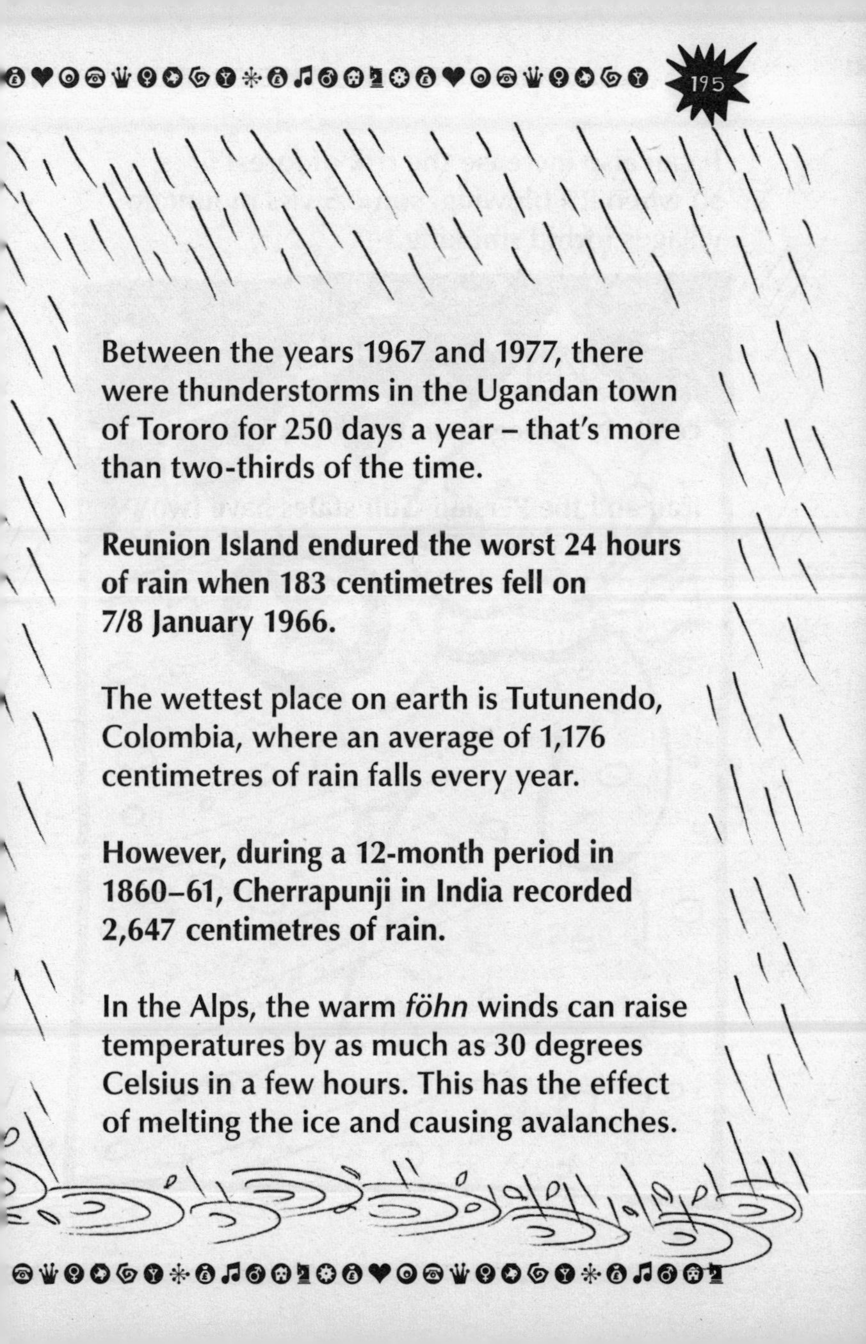

Between the years 1967 and 1977, there were thunderstorms in the Ugandan town of Tororo for 250 days a year – that's more than two-thirds of the time.

Reunion Island endured the worst 24 hours of rain when 183 centimetres fell on 7/8 January 1966.

The wettest place on earth is Tutunendo, Colombia, where an average of 1,176 centimetres of rain falls every year.

However, during a 12-month period in 1860–61, Cherrapunji in India recorded 2,647 centimetres of rain.

In the Alps, the warm *föhn* winds can raise temperatures by as much as 30 degrees Celsius in a few hours. This has the effect of melting the ice and causing avalanches.

It can also increase the risk of forest fires, so when it's blowing, some Swiss mountain villages forbid smoking.

In Seville in Spain on 4 August 1881, the temperature reached 50 degrees Celsius – the hottest in the history of Europe.

The highest temperature ever recorded anywhere in the world was in Al 'Aziziyah, Libya, on 13 September 1922, when the thermometer climbed to a mind-boggling 57.8 degrees Celsius.

Brunei's lowest recorded temperature is 21 degrees Celsius.

The way other people live (2)

Places of entertainment such as theatres and cinemas are banned in Saudi Arabia.

In Mexico, a left-turn signal on the vehicle in front of you could mean the driver is turning left or it could mean that it is clear ahead and you can overtake.

In Poland, most people celebrate their name days rather than their birthdays. A person's name day is the day of the year dedicated to the particular saint after whom they are named.

Many Chinese people wear jade for protection. It is believed to bring good luck and good health.

In Afghanistan, people are careful to ensure that the soles of their feet don't point at another person. Slouching or stretching your legs in a group is offensive. Men and women try not to make eye contact during conversation and they also don't display

affection in public – even if they're married.

Every year the Chinese use 45 billion chopsticks.

On weekends, the people of Qatar enjoy driving to oases, where they camp out in tents. For a night or two they experience the lifestyle of their Bedouin ancestors.

Bolivians maintain little personal space and tend to stand close during conversations.

Elephants are an important part of Sri Lankan culture. They are decorated for religious processions and their images appear in temples and palaces. Some elephants are still used for work. Others are in danger of extinction due to uncontrolled expansion of agriculture and development.

The United Arab Emirates' largest city – Dubai – has become a haven for tourists from around the globe. Here are just five of its incredible attractions:

UAE Spaceport: a terminal for tourists going into – and returning from – space.
Burj Khalifa: set to be the world's tallest

building at 960 metres.
Hydropolis: an underwater hotel resort.
The Palm Islands: artificial islands that are shaped like a massive palm tree.
The World: 250 artificial islands that form an elaborate map of the world, which can be seen from space on clear days.

1 January is every Korean's birthday. On this day they add another year to their age, irrespective of their actual date of birth. Birthdays are special celebrations, particularly the first and the 60th ones.

December is the most popular month for weddings in the Philippines.

Ethiopians usually don't celebrate their birthdays. Dates of birth are registered only in some urban areas, and many Ethiopians aren't even interested in which day of the year they were born.

An old Swedish law called 'Every Man's Right' means that it is legal to visit somebody else's land, to bathe in their lake, travel by boat on their river, and pick the wild flowers,

berries and nuts found there.

In the Iranian city of Yazd – the second most ancient and historic city in the world – houses are built below ground to escape the heat. These houses have special 'wind towers' designed to catch the wind and send it down to refresh the living areas.

Because the traffic moves so very slowly along the congested roads of Bangkok, many parents leave home with their sleeping children early in the morning. They feed and dress their children when they actually get to school.

For the Japanese the stomach is the centre of the emotions. Instead of having heart-to-heart talks, the Japanese say they 'open their stomachs' for a good conversation.

The concept of privacy is alien to Jordanians; the closest word in Arabic translates into the English word 'loneliness'. An old proverb says, 'Where there are no people, there is hell.'

In Rwanda, farmers used to decorate the walls of their buildings with cow dung, which they sculpted and painted in bold geometric designs.

Largest

Russia is the largest country in the world by far – it has over 17 million square kilometres, which is almost double that of the next biggest countries which are Canada, China and the US, all of which are over nine million square kilometres (the UK has a mere 244,820 square kilometres). Russia stretches over two continents (Europe and Asia) and 11 time zones.

Nicaragua is the largest country in Central America.

Colca Canyon, a canyon of the Colca River in Peru, is the world's largest canyon. It's twice as deep as the Grand Canyon (though its walls are not as vertical). The name *Colca* refers to small holes in the cliffs in the valley and canyon. These holes were used in Inca and pre-Inca times to store food, such as potatoes. They were also used as tombs for important people. The Colca Canyon is home to the Andean condor.

Excluding Antarctica, the world's largest desert is the Sahara in North Africa. This is followed by the Arabian, Gobi and Patagonian. There are no deserts in Europe.

With 72 million books, the Library of Congress in Washington, DC, US is the largest library in the world.

The US also boasts the largest amusement park in the world: Orlando's Disney World in Florida (which covers 121,400 square metres).

The largest employer in the world is the Indian railway system, which employs over a million people. Indian Railways transports more than six billion passengers a year. Every day, it runs over 14,000 trains.

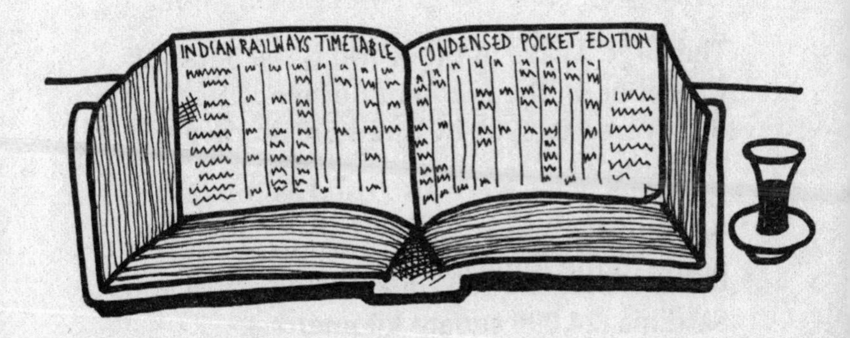

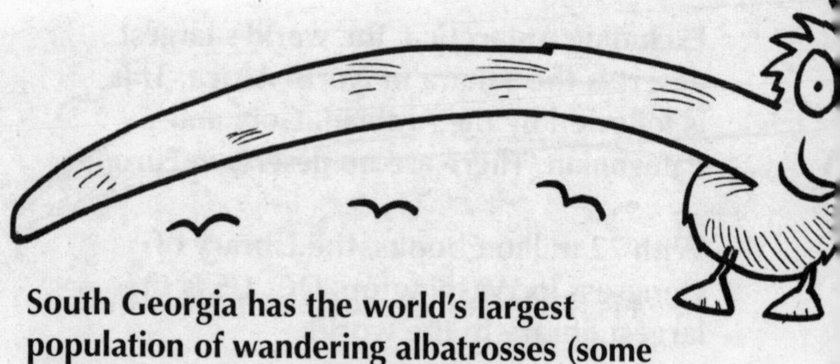

South Georgia has the world's largest population of wandering albatrosses (some 4,000). At 3.6 metres, the wandering albatross has the largest wingspan of any bird in the world.

At 2,166,086 square kilometres, Greenland is the world's largest island. Despite this, it has a tiny population of fewer than 60,000 people. That's because it's also more than four-fifths ice-capped. It is, however, home to the world's largest national park.

The largest cave in the world is the Sarawak Chamber on the island of Borneo. It was discovered in 1981 by three British cave explorers.

The two largest islands in Europe are both Italian: Sicily (25,460 square kilometres) and Sardinia (24,090 square kilometres).

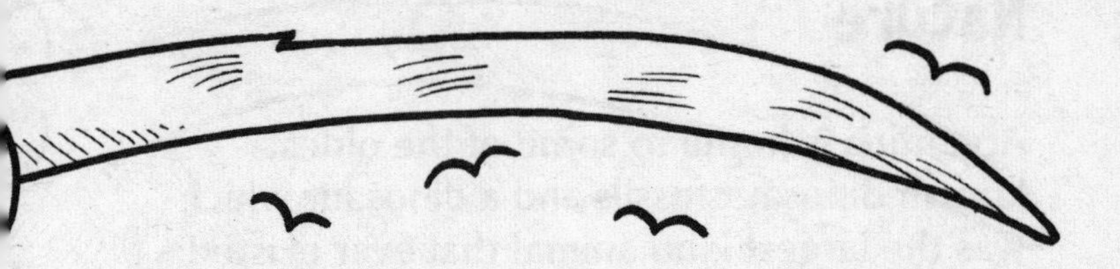

India is the largest democracy in the world.

At over one and a half kilometres, Victoria Falls, on the border of Zambia and Zimbabwe – is the world's largest waterfall. It is a World Heritage Site, is twice as high as Niagara Falls and one and a half times as wide, and is said to be the longest curtain of water on Earth. The sound of crashing water emanating from Victoria Falls is so loud that it can be heard from 50 kilometres away.

The largest city in Africa is Cairo in Egypt.

With an area of 2,505,813 square kilometres, Sudan is the largest country in Africa.

The Pacific Ocean is the world's largest ocean. It covers 28 per cent of the Earth's surface and contains 46 per cent of the world's water.

Nature

Argentina is home to some of the oldest known dinosaur fossils and a dinosaur which was the largest land animal that ever existed: the Argentinosaurus.

Go along the Selangor River near Kuala Lumpur, the capital of Malaysia, at night and you'll see thousands of fireflies flash on and off like Christmas lights. Scientists can't explain this phenomenon. Resourceful villagers used to put the fireflies in bottles and use them as lamps.

There are giant bats in Indonesia with a wingspan of nearly 1.8 metres.

There's a species of earthworm in Australia that grows up to three metres in length.

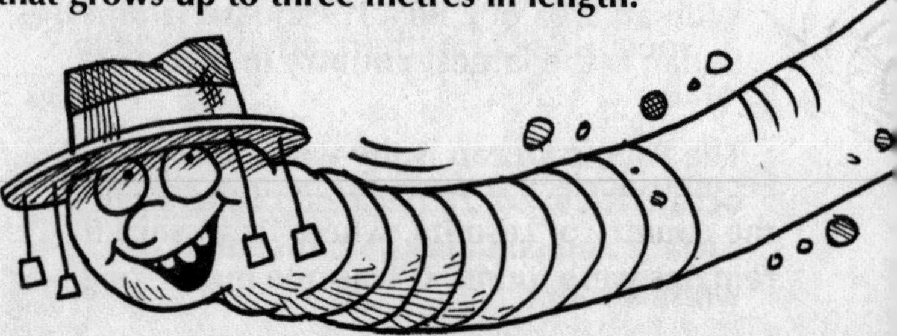

Libya has extremes of climate that range all the way from intense heat to severe frosts. Found near the sea or by desert oases, the palm tree thrives in these extreme conditions. This is fortunate for the nomads who use the trunk for fuel and for making rope, and weave the leaves into sandals and baskets.

Peru is home to more than 1,800 bird species, 120 of which are found nowhere else in the world.

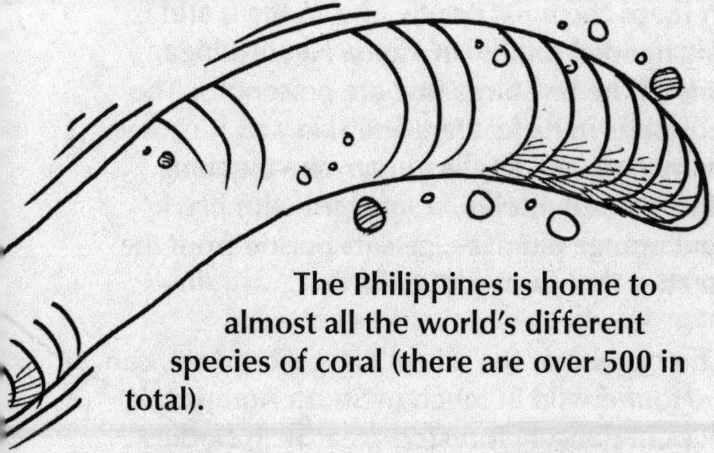

The Philippines is home to almost all the world's different species of coral (there are over 500 in total).

Lesothosaurus is a dinosaur named after the country of Lesotho, where its fossilized remains were found. It was one metre long

and lived around 200 million years ago. In fact, dinosaur footprints can be found in many places in Lesotho.

Komodo dragons are named after the Indonesian island of Komodo where they can still be found. They're not dragons but lizards; they're about three metres long and weigh up to 140 kilograms. They have scaly bodies, short muscular legs, massive tails and razor-sharp teeth.

Perhaps the most deadly bird in the world is the hooded pitohui of Papua New Guinea, one of the few birds that are poisonous. The poison is in its feathers and skin and it causes numbness and tingling in anyone touching. The hooded pitohui, a songbird with black and orange plumage, gets its poison from the beetles that form part of its diet.

The capybara, the world's largest rodent, can be found wild in much of South America. It lives in densely forested areas near bodies of water (it is semi-aquatic). The only time they're found further afield is when they escape from captivity – which is why one was found in the River Arno in Florence, Italy, in 2008!

The more lush the vegetation, the more poisonous the toads that live in it. Scientists have found that toads in Madagascar deliberately eat insects high in poisonous alkaloids that then ooze from the toads' skin glands. If these toxic creatures are fed different foods – for example in captivity – they cease to be poisonous.

Food & drink

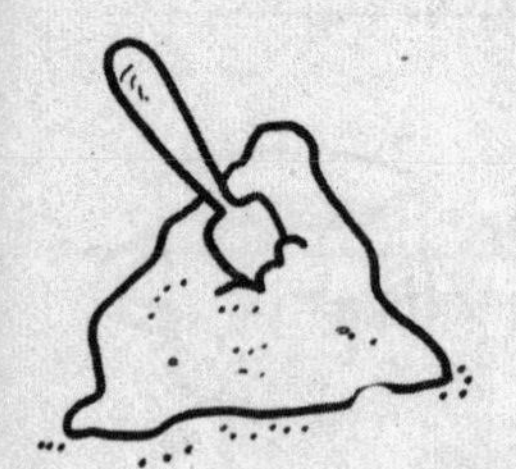

Syria is one of the highest consumers of sugar, with soft drinks that are the sweetest in the world.

Many Hondurans grow pineapples in their gardens. Every part of the pineapple is used for something. The skin is used to make tea or vinegar for preserving vegetables. The fruit is used to make juice, jam or pies. The tops are put in buckets of water until they sprout roots and can be replanted in the garden.

More than 99 per cent of the world's food supply comes from the land; less than one percent comes from the oceans or rivers.

In Germany, there are 5,000 different types of beer and 300 different types of bread.

Tea is the national drink of Afghanistan.

Mali is famous for its salt mines. At one time, salt was such a valuable commodity that people would trade a pound of gold for a pound of salt.

In Mexico, chocolate, which comes from the bean of the cacao tree, was known as the drink of the gods, because by law only the nobility could drink it.

The Danish eat the most frozen food.

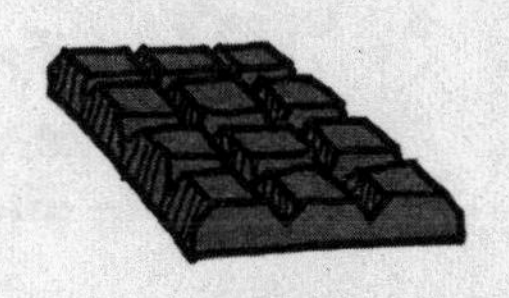

Ukraine used to be known as the bread basket of Europe – because of its wheat fields.

Doughnuts originated in Holland.

In France, the average person drinks over 25 gallons of wine per year.

In Iceland, hot dogs are usually made with lamb rather than beef or pork.

Clementine oranges are originally from west Algeria. They are named after a French priest named Father Clément, who found them growing in the village of Misserghin.

In Somalia, some nomads drink a fermented beverage which is made by burying camel's milk in a leather flask for a week.

Traditionally, the Malagasy people of Madagascar serve food on a mat on the floor.

Cinema snacks

Popcorn and sweets are the most popular cinema snacks in Britain but elsewhere in the world they eat other things.

In China they prefer salted plums, or chunks of tofu in a spicy sauce.
In Japan they like fried squid pancakes or octopus kebabs.
In Indonesia they eat dried fish and cuttlefish.
In Korea they like to chew on roasted octopus tentacles – including the suckers.

The traditional thing to eat on Christmas Day in Greenland is seabirds wrapped in the skin of a seal.

In Japan there is a liquor made from the fermented bodies of venomous snakes.

Chicle, a milky sap extracted from a tree found in Mexico, is the vital ingredient in chewing gum.

The Filipinos make complete use of the

coconut tree. They turn the fruit into copra (dried coconut flesh) and oil, and use the tree's leaves to make hats and balls. Meanwhile, the trees themselves are used for construction.

One day in Poland, a brewery developed a plumbing problem in which beer was accidentally pumped into the incoming water supply. The result was that residents of the town got free beer on tap in their kitchen sinks and bathrooms.

There is no refrigeration in most of Laos, so all meals must be freshly prepared. Mint is often used to keep food fresh.

More chocolate is sold at Brussels Airport than anywhere else in the world.

In Laos, people eat giant water bugs.

Potatoes have been a staple of the Irish diet for hundreds of years. Before the potato famine of the 19th century, the average person ate six kilograms of potatoes a day. Even today, the Irish still eat more potatoes than the people of any other European country.

The average French person eats 800 grams of garlic a year.

For Australia's Aborigines, an insect larva known as a 'witchetty grub' is a delicacy. Found at the roots of the acacia plant, the wichetty grub is white and the size of a baby carrot.

The durian is the smelliest fruit in the world. In fact, it's so smelly that in those countries to which it's native (Brunei, Indonesia and Malaysia) there are markets and restaurants which have special rooms for people who want to enjoy the fruit without its smell disturbing other people.

In Cyprus, salt is collected from the salt lakes in summer, when the heat dries up the water, exposing the salt.

Argentinians eat more meat than any other people.

The sugar cube was invented in (what is now) the Czech Republic in 1843.

Mocha was named after the Yemeni town of Mocha. It was here that the coffee bean with the hint of chocolate was first discovered.

France has more than 300 different kinds of cheese.

Mongolians put salt in their tea instead of sugar.

Health

In 1976 doctors in Los Angeles went on strike because of the rising cost of malpractice insurance. All non-emergency surgery and medical attention were cancelled. During the strike, there were 18 per cent *fewer* deaths than usual.

In Somalia, sheep fat is used to treat a variety of ailments, from rheumatism to broken bones and chest pains.

Malaysia's national flower, the hibiscus, is also used as a herbal remedy for many complaints.

Rather than going to a doctor or hospital, many Eritreans prefer taking a dip in one of the country's hot springs, which are believed to cure sickness.

Iranians use the spice saffron to help relieve digestive and nervous disorders. The country has produced and exported saffron since ancient times. It takes more than 1,000 crocus flowers to produce less than 15 grams of this extremely expensive spice.

Nepal and Benin share the dubious distinction of having the fewest hospitals per person in the world: just three for every 10,000 people.

Monaco, on the other hand, has the most hospitals and the most hospital beds per person, with 163 for every 10,000 people.

Kava, Fiji's most popular drink, is said to have medicinal properties and is used by Fijians to cure a whole range of ills. Even though it's not alcoholic, it still relaxes you. However, I have to tell you that it tastes vile!

Some Belarusians treat colds by drinking vodka spiked with salt and pepper, or milk with an egg yolk and honey.

Pyrethrum, which grows in Rwanda, is a

daisy-like flower used to make a natural insecticide. Its properties were discovered during the First World War, when a group of soldiers camped overnight in a field of pyrethrum. By morning, the lice that had infested the soldiers had all been killed.

The Vietnamese use the traditional remedy of cupping. A small glass cup is heated and the open end is placed on the skin, causing small blood vessels to break. The treatment is believed to draw 'poisonous wind' out of the system.

In Bulgaria, fire-dancing, an ancient religious and ritual dance on burning embers, was believed to get rid of illness and restore good health.

In 1811 two boys joined at the chest were born in a village near Bangkok in what was then known as Siam but is now called Thailand. They travelled the world and became celebrities, but they were never able to find a surgeon who could separate them. Since then, twins who are joined at birth have been called Siamese twins.

In 1991 a mummified body from the Stone Age was found in the ice of the Ötztal Alps between Austria and Italy. Tattoos, thought to have been for medicinal purposes (perhaps a form of acupuncture), were found on the body.

Highest

The world's highest cricket ground is in Chail, Himachal Pradesh. Built in 1893 after levelling a hilltop, the pitch is 2,444 metres above sea level.

The world's highest railway is in Peru. It's used to transport people to the lost Inca city of Machu Picchu.

The sand dunes in the Namib Desert are the highest in the world.

At 8,848 metres, Mount Everest is, of course, the highest mountain in the world. It's in the Himalayas as are many of the world's highest mountains. The Himalayan mountains are 70 million years old.

Unsurprisingly, the Everest View Hotel – with an altitude of just under four kilometres – is the world's highest hotel.

Mount Kilimanjaro is the world's biggest free-standing mountain – that's to say, it's

the biggest mountain that's not part of a mountain range.

The highest home in the world is a shepherd's hut in the Andes (5,180 metres above sea level).

Mexico City is North America's highest city.

The world's highest dam is the Nurek in Tajikistan. It's 300 metres high. However, it's expected to be dwarfed by the new Rogun Dam, which has a planned height of 335 metres. And where is the Rogun Dam going to be built? Yes, you guessed it – Tajikistan!

Nepal and China are the highest countries in the world.

Chile has the highest volcano in the world (Ojos del Salado, which is 6,891 metres high).

With a mountain 120 metres above sea level and a lake 40 metres below sea level, the Dominican Republic has both the highest and lowest points in the Caribbean.

Lowest

Iceland has the lowest murder rate and also the highest reported crime rate.

Belgium has the lowest number of McDonald's restaurants per person in Europe.

At an average depth of -4,300 metres, the Pacific Ocean is the deepest ocean in the world.

The bottom of Lake Tanganyika, at 358 metres below sea level, is the lowest point in Africa. This lake lies in the Great Rift Valley, a geological fault line that extends for 9,700 kilometres, crossing the continent from Jordan to Mozambique.

History (2)

The early 18th century outlaw Juraj Jánošík is a national folk hero of Slovakia. Like Robin Hood, he avenged injustices and stole from the rich to give to the poor. He has inspired more artistic works than any other personality in Slovakia's history. In Slovakia, Robin Hood is known as the English Juraj Jánošík!

During the reign of Peter the Great, Russian noblemen weren't allowed to grow beards unless they paid the Beard Tax. Peter the Great wasn't just trying to raise money though: he was also jealous of men with beards as he couldn't grow much of a beard himself!

The ancient Aztec Indians of Mexico were great warriors. They believed that if they made human sacrifices to their gods, they would be blessed with good crops, fine weather, and victory in war.

In 1778 Prussia and Austria fought the Potato War in which each side tried to starve the other by consuming their potato crop.

In the 17th century a craze called 'tulipomania' swept the Netherlands. Tulips became so popular that people had to pay huge sums to buy a single bulb. Eventually, the government regulated the tulip trade.

In 1830 King Louis XIX ruled France for 15 minutes.

The Hanging Gardens of Babylon, one of the Seven Wonders of the World, were built in Iraq around 600 BC by King Nebuchadnezzar II for his wife, Amytis, to help her get over her homesickness for mountainous northern Persia (Iran).

The imperial throne of Japan has been occupied by the same family for the last 1,300 years.

Genghis Khan was emperor of the vast Mongol Empire, the biggest joined-up empire in history. It covered most of Asia and stretched as far as Hungary.

In 1765, the British government paid £70,000 for the Isle of Man.

Over 30,000 werewolf cases were tried in France between 1520 and 1630.

During the 17th century pirates buried stolen treasure on Isla de la Juventud (Isle of Youth), off Cuba. According to legend, this was the

famous Treasure Island described in Robert Louis Stevenson's novel.

The Taj Mahal was built by Emperor Shah Jahan in memory of his queen, Mumtaz Mahal, at Agra, India. Mumtaz and Shah Jehan were married in 1612 and had 14 children together. In 1630, she gave birth to her last child and died soon after. So great was Shah Jahan's love for his late wife that he built the most beautiful mausoleum on Earth for her. About 20,000 people worked on its construction – entirely in marble. The Taj Mahal is the world's most symmetrical structure, with the empress's tomb exactly in the centre. The gardens, number of slabs of marble, engravings of flowers, even the number of leaves carved and laden with emeralds on each side, are all perfectly symmetrical.

Children's teeth

They have a first-tooth ceremony in Spain. Whoever finds the first tooth a child loses becomes that child's godmother. Obviously, parents take great care to make sure that the right person finds the tooth!

In Malaysia, when a child loses a tooth, they bury it because it is a part of their body and needs to be returned to the Earth.

When Afghan children lose a tooth, they drop it inside a mouse hole, saying, 'Take my dirty old tooth and give me your small clean one instead.'

In Cambodia, if a child loses a tooth, they throw it on the roof. If they lose an upper tooth, they put it under their bed. Their parents tell them that the new tooth will grow towards the old one and come in straight.

In Botswana, children throw their tooth

onto the roof and say, 'Mr. Moon, Mr. Moon, please bring me a new tooth.'

In Haiti, children throw the tooth onto the roof and say, 'Rat, Rat, Rat. I give you a beautiful tooth. Send me back an old tooth.' They say the opposite of what they mean in order to trick the crafty rat into giving them what they really want.

When Libyan children lose a tooth they throw it to the sun and say, 'Bring me a new tooth.' They can then be sure to have a bright smile because their teeth came from the sun.

When Argentinian children lose a tooth they put it in a glass of water. During the night a little mouse called El Ratoncito comes and drinks all the water, takes the tooth, and leaves some coins or sweets in the empty glass.

In Tajikistan, a boy's milk teeth are 'sown' in the fields so that he'll grow up to be a warrior.

Language (2)

Polish has some lovely names for months. For example, April in Polish means 'flowers', July means 'linden tree', September means 'heather' and November translates as 'falling leaves'.

The way that Australians speak is known as 'strine' – thought to be the word 'Australian' spoken through closed teeth. Some scholars say this pronunciation came about because of the need to keep one's mouth shut against blow flies.

The language of Manx is taught in almost every school on the Isle of Man.

Between the 17th and 19th centuries, French was the language of diplomacy and culture in Europe, and many northern European monarchs spoke it in their courts. Today, there are some 125 million French speakers worldwide.

Jamaicans refer to their major crops in terms

of gold. Sugar is brown gold, bananas are green gold and citrus fruits are sun gold.

Over 800 different languages are spoken in Papua New Guinea. With a population of just over six million people, that's about one language for every 7,500 people. The result of this is that there are villages within a few kilometres of each other which speak different languages.

Asia has the most languages and the most speakers, accounting for 61 per cent of all language speakers in the world.

In the Netherlands, they have a wonderful expression of amazement that translates as 'That breaks my clog'. The sturdy wooden clog is still worn by some Dutch farmers because it is waterproof in damp fields.

Rather than merely saying 'goodbye', an Irish person might say 'safe home'. Another common expression is 'God bless the work', a greeting used when entering the company of a person who is working. When the Irish say something is 'great craic' (pronounced 'crack'), they mean they're having fun.

Luxembourg has its own language – Luxembourgish – which is also sometimes spoken in parts of Belgium, France and Germany.

In the northern province of Friesland in the Netherlands, the children learn Frisian, the local language, as well as Dutch and English.

The Filipino language reflects the importance of rice to the Filipino people. Their language has several words for rice – including words to cover rice that is harvested but not cleaned, rice that is still cooking in a pot and rice that is ready to eat.

Disney around the world

Walt Disney wanted to build a park for his employees and their families near his Burbank studios but, over time, his dream grew and so did his plans. He bought over 160 acres of orange groves around Anaheim and, in 1954, set about building his 'Magic Kingdom'.

Originally he planned a nine-million dollar 45-acre park, but by the grand opening day, 17 July, 1955, the park covered the full 160 acres. After the success of Disneyland California, Walt Disney and his brother, Roy, began buying land near Orlando, Florida – eventually ending up with 27,000 acres. Disney World, the Magic Kingdom, was much bigger than Disneyland. It was the first of Disney's many theme parks in Florida.

There are currently five Disney resorts in the world:

Disneyland Los Angeles, 1955
Disney World Florida, 1971

Disneyland Paris, 1992*
Disneyland Tokyo, 1983
Disneyland Hong Kong, 2005

There's also a community in Florida named Celebration: a town where people actually live all year round. World Drive connects Celebration directly to the Walt Disney World parks and resorts; the north end of World Drive begins near the Magic Kingdom and its south end connects to Celebration Boulevard, allowing Celebration residents and guests to drive to Disney attractions without having to use any busy roads.

*Not all French people wanted Disneyland Paris to be opened, to the extent that some outraged citizens took their anger out on the park's characters and started beating up Mickey and Goofy.

Genuine products from overseas

Mukki (Italian yoghurt)
Plopp (Swedish toffee bar)
Bum (Turkish biscuits)
Zit (Greek fizzy drink)
I'm So Sorry Please Forgive Me (Swiss chocolate bar)
Kevin (French aftershave)
Meltykiss (Japanese chocolate)
Happy (Swedish chocolate)
Noisy (French butter)

Longest

The Nile is the world's longest river.

Australia's Great Barrier Reef is the longest coral reef in the world.

At a staggering 1,896 kilometres, Yonge Street in Canada (starting at Lake Ontario, splitting Toronto into east and west, finally ending in Rainy River) is the longest street in the world.

Who can say how long the longest fart is? For all we know, someone is letting off a record-breaker right at this minute – in complete privacy! However, a London man named Bernard Clemmens managed to sustain a fart for an officially recorded time of two minutes, 42 seconds.

Chile is the world longest country in relation to its width. It is over 20 times longer than it is broad. It stretches over 4,250 kilometres from north to south but it is never more than 500 kilometres wide.

The Andes mountain range runs along the entire length of the country like a backbone.

A Norwegian man, Hans Langseth, had the longest beard in history. By the time of his death in 1927, his beard measured over five metres. That's over three times the height of the average person. The beard is now on display in the Smithsonian Institution in the US.

Music

In Angola, the *saxi* is an instrument made from dried fruit filled with dried seeds or glass beads. It's a percussive instrument rather like the maracas.

In Zambia, the *vingwengwe* is an instrument played by four women. Four overturned metal pots are placed in a row. Each woman places a stool on top of the pot and then turns it to make the pots resonate. As they turn the stools, they create a quartet of 'voices'.

The *rebaba* is a traditional musical instrument used in Arab countries. It has just two strings (tuned to C and G) with a four-octave range.

Flamenco is Spanish dance music at its most powerful. It's a style of music where song, dance and guitar are blended into rhythms that are often improvised (i.e. made up on the spot). Often, the tragic lyrics and tone reflect the sufferings of the gypsy people with whom it originated. However, nobody really

knows the origin of the word itself as the word means either Flemish or flamingo – and neither seems appropriate to flamenco!

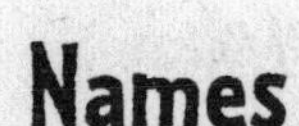

Names

Many Bosnian names end in 'ic', which means 'child of'.

Up until fairly recently, the people of Outer Mongolia didn't have surnames.

In the past, Laotians used only their first names. They started using family names in 1943, when a law was passed making the use of surnames compulsory. However, Laotians sometimes change their names when they start a new job or move to a new place – or even just to mark a turning point in their lives. Perhaps the most common reason for changing names is to ward off evil spirits.

In Somalia, people have three names: their first name, followed by their father's and grandfather's names. Because of the clan system, many people have very similar names.

In China, a population of over a billion shares just 200 family names.

Many Irish family names start with 'Mc' or 'O', which mean (respectively) 'son of' and 'grandson of' in Gaelic.

Swedes didn't have family names until about 100 years ago. Before then the suffixes 'son' or 'dotter' were attached to the father's name when referring to a boy or a girl. So if, for example, a man named Lars had a son named Stefan, he would be called Stefan Larsson. On a similar basis, his daughter Gerda would be called Gerda Larsdotter.

Traditionally, Ethiopians don't have last names. Instead, they're given a personal name at birth. Their second name is their father's personal name. If they choose to use a third name, it will be the personal name of their paternal grandfather (i.e. their father's father).

In Kazakhstan, people won't name a child after a relative as it implies that the child is replacing that person.

Idi Amin, the tyrant of Uganda in the 1970s, gave himself the title: His Excellency,

President for Life, Field Marshal Al Hadji Doctor Idi Amin Dada, Lord of All the Beasts of the Earth and Fishes of the Seas and Conqueror of the British Empire in Africa in General and Uganda in Particular.

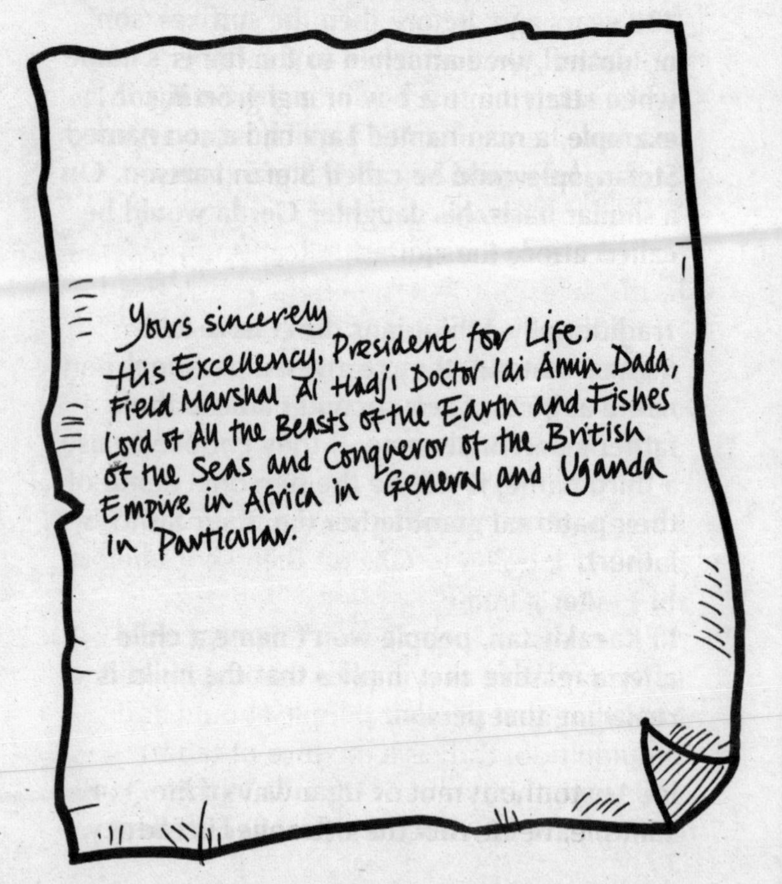

Yours sincerely
His Excellency, President for Life,
Field Marshal Al Hadji Doctor Idi Amin Dada,
Lord of All the Beasts of the Earth and Fishes
of the Seas and Conqueror of the British
Empire in Africa in General and Uganda
in Particular.

Customs & traditions

Presenting bread and salt to a visitor is a traditional form of hospitality in Belarus.

In Burundi, the importance of cattle is demonstrated in the traditional greeting, 'May you have herds of cattle', which is a way to wish people good health and prosperity.

Scarification, the ritual application of scars to the faces of tribal members, is practised by many groups in Chad. Lines and other symbols are made on the faces of young men to mark them permanently as members of a particular tribe.

Koreans do not wear shoes in the house. They consider it polite to take off their shoes before they enter a home.

In Madagascar, they have a set of rules called Fady that govern what people should and shouldn't do. Fady is a mixture of taboo and custom but mostly it's a way of life. For example, being rude to a stranger is Fady;

singing while you're eating (apparently, it will lead to elongated teeth) is Fady; having a funeral on a Tuesday is Fady (it could lead to another death in the family); children eating their meals before their elders eat theirs is Fady. Sometimes, what is Fady can be hard for outsiders to follow: it may be particular to a certain region of the country. For example, visitors to the archaeological site of Ambohimanga must enter the site with their right foot first and leave with their left foot first.

Very young Burmese children wear holy thread around their necks or wrists to protect them from bad spirits or spells.

After a wedding, Somalis hold dances for three to seven days in a row.

In Malaysia, they celebrate a birth when the child is one month old . . . by shaving the child's head.

Although Thais like to give and receive presents (as we all do!), it's considered rude to open a gift in the presence of the giver. So

they put it aside and only open it when they're alone.

When people in Mauritius are given a present, they always accept it with their right hand.

In Korea, it's considered good manners to give and receive objects with both hands. If Koreans want to receive a gift, they'll put out their right hand to accept it and their left hand under the right wrist.

If a Congolese offers to share a meal with a guest, that guest is expected to show reluctance to join the host's table. But the guest should ultimately accept the offer. Not to do so would be considered rude.

It is customary for Rwandan children to avoid eye contact when talking to an elder.

Zambians pride themselves on their hospitality. There's a Zambian proverb that says, 'Do not look at a visitor's face but at his stomach.'

Every year there's a Great Tomato Fight in Buñol, Spain. Tens of thousands of people come from all over the world to participate in a great battle, during which more than a hundred tons of over-ripe tomatoes are

thrown in the streets. For the week leading up to the battle, the town of Buñol is filled with parades, fireworks, food and street parties.

In Rwanda, rich women often wore heavy copper bracelets and anklets. Because of the weight of the jewellery, the women were unable to do much work. So the very wearing of jewellery served to distinguish rich women from those who worked in the fields.

Cows are considered sacred in India so they're free to roam the city streets without being stopped by people. In New Delhi, microchips are sometimes put into the stomachs of cows so that, if they got lost, they can be traced and returned to their owners.

Hairdressing was once a traditional art form in Fiji, where a man's hair expressed his power and social status. Fijian men dyed their hair various colours and styled it into fantastic shapes, with chiefs' hairdressers taking days to create special effects.

Men from the Dassanech tribe of Ethiopia daub their bodies with cow dung.

In Thailand, when two strangers meet, they begin by establishing who has the higher status. Questions such as 'How old are you?' and 'How much do you earn?' are not considered rude.

In Thailand, it's a sign of poor manners to point at someone with your feet.

In Japan, it is customary to wash before getting into the bath. The bath is for soaking and relaxing, not for washing. The water stays clean so that another family member can soak in the same water later.

Afghans traditionally celebrate the birth of a

child on its sixth day. On this day the baby is given a name, and guests visit with gifts.

Young Croatians celebrate the arrival of the summer solstice by jumping over bonfires.

At Czech weddings they throw peas instead of rice.

The Berbers of Algeria pride themselves on their hospitality. They have a saying: 'When you come to our house, it is we who are your guests, for this is your house.'

In Algeria, using your fingers to point at objects or people is considered impolite. Care is also taken not to let the sole of the foot point at another person.

In the Andes, before people drink chicha, a maize-based drink, they sprinkle a few drops onto the ground for Pachamama (the goddess known as 'Mother Universe') to 'guarantee' a good harvest.

In many Asian cultures, the head is the most sacred part of the body – so touching a person's head is taboo.

In Tanzania, every year a torch is lit on Mount Kilimanjaro and then carried across the country by runners to celebrate the country's independence.

That's surprising! (4)

In Guatemala, lemons are green and limes are yellow.

There was once a devout Hindu who, to demonstrate his mastery over pain, raised his arm above his head and kept it that way for years. He held his arm so still that a bird built a nest on his hand.

Hailstones – each weighing approximately one kilogram – fell in Bangladesh in 1986

Nepal has an airline company called Buddha Air.

There's a street in South Africa that is home to two winners of the Nobel Prize for Peace: Nelson Mandela and Desmond Tutu.

At their nearest point, Russia and America are less than four kilometres apart.

Although the Maldives are made up of over 1,000 islands, the combined land mass only amounts to 300 square kilometres.

Lasts

In 1999 Bhutan became the last country in the world to introduce television.

Luxembourg is the last surviving grand duchy – that's to say it's the only country in the world that still has a grand duke.

The town of Sauteurs ('leapers' in French) in Grenada is so named because it is believed that the last Carib warriors jumped to their death from a nearby cliff, rather than surrender to the Europeans.

Mauritius was the home of the dodo until it became extinct. The last dodo died there in 1681.

The guillotine was last used in France publicly in 1939 and behind prison walls in 1977.

Mitchell Symons

WHY EATING BOGEYS IS GOOD FOR YOU

. . . and other crazy facts explained!

Ever wondered . . .

- **Why we have tonsils?**
- **Is there any cream in cream crackers?**
- **What's the best way to cure hiccups?**
- **And if kangaroos keep their babies in their pouches, what happens to all the poo?**

Mitchell Symons answers all these wacky questions and plenty more in a wonderfully addictive book!

(And eating bogeys *is* good for you . . . but only your own!)

Selected for the Booktrust Booked Up! Initiative in 2008

Mitchell Symons

HOW MUCH POO DOES AN ELEPHANT DO?*

... and further fascinating facts!

* An elephant produces an eye-wateringly pongy 20 kilograms of dung a day!

Let Mitchell Symons be your guide into the weird and wonderful world of trivia.

- Camels are born without humps
- Walt Disney, creator of Mickey Mouse, was scared of mice
- Only 30% of humans can flare their nostrils
- A group of twelve or more cows is called a flink

Mitchell Symons

WHY DO FARTS SMELL LIKE ROTTEN EGGS?

... and more crazy facts explained!

Ever wondered . . .

- **Why we burp?**
- **What a wotsit is?**
- **Whether lemmings really jump off cliffs?**
- **Why vomit always contains carrots?**
- **And why *do* farts smell like rotten eggs?**

No subject is too strange and no trivia too tough for Mitchell Symons, who has the answers to these crazy questions, and many more.